NASA Reference Publication 1344

1994

Total Solar Eclipse of 1995 October 24

Fred Espenak
Goddard Space Flight Center
Greenbelt, Maryland

Jay Anderson
Environment Canada
Winnipeg, Manitoba
CANADA

National Aeronautics and Space Administration

Scientific and Technical Information Branch

PREFACE

This is the third in a series of NASA Eclipse Bulletins containing detailed predictions, maps and meteorological data for future central solar eclipses of interest. Published as part of NASA's Reference Publication (RP) series, the bulletins are prepared in cooperation with the Working Group on Eclipses of the International Astronomical Union and are provided as a public service to both the professional and lay communities, including educators and the media. In order to allow a reasonable lead time for planning purposes, subsequent bulletins will be published 24 months or more before each event. A tentative schedule for future eclipse bulletins and projected publication dates appears at the end of the Preface.

Response to the first two eclipse bulletins was overwhelming. When the January 1994 issue of *Sky and Telescope* announced their availability, as many as 160 requests per week were received for them. Since requests for the May bulletin outnumbered the November bulletin by four to one, an additional 600 copies of RP1301 were reprinted in late January. Nevertheless, the demand quickly exceeded the supply for both bulletins and funding sources did not permit more printings to fill all requests. It also became clear that the requests were consuming a great deal of time, secretarial work and postage. To conserve resources and to make responses faster and more efficient, the proceedure for requesting eclipse bulletins has been modified as follows.

Single copies of the bulletins are available at no cost and may be ordered by sending a 9 x 12 inch SASE (self addressed stamped envelope) with sufficient postage (11 oz. or 310 g.). Use stamps only; cash or checks cannot be accepted. Requests within the U. S. may use the Postal Service's Priority Mail for $2.90. Please print either the NASA RP number or the eclipse date (year & month) of the bulletin ordered in the lower left corner of the SASE. Requests from outside the U.S. and Canada may use international postal coupons sufficient to cover postage. Exceptions to the postage requirements will be made for international requests where political or economic restraints prevent the transfer of funds to other countries. Finally, all requests should be accompanied by a copy of the request form on page 73. Bulletin requests may be made to either of the authors. Comments, suggestions, criticisms and corrections are solicited to improve the content and layout in subsequent editions of this publication series, and may be sent to Espenak.

Since we are now entering the age of the 'Information Highway', it seems fitting that the eclipse bulletins should be served electronically. Thanks to the initiative and expertise of Dr. Joe Gurman (GSFC/Solar Physics Branch), the first three eclipse bulletins are all available over the Internet. Formats include a BinHex-encoded version of the original MS Word file + PICT + GIF (scanned GNC maps), as well as a hypertext version. They can be read or downloaded via the World-Wide Web server with a mosaic client from the SDAC (Solar Data Analysis Center) home page: *http://umbra.gsfc.nasa.gov/sdac.html*. Most of the files are also available via anonymous ftp. In addition, path data for all central eclipses through the year 2000 are available via *http://umbra.gsfc.nasa.gov/eclipse/predictions/eclipse-paths.html*. For more details, please see pages 17 and 18. Naturally, all future eclipse bulletins will also be available via Internet.

RP 1344 marks a milestone as the first eclipse bulletin to be generated entirely on a Macintosh computer (excluding the GNC maps). As such, it represents the culmination of a two year project to migrate a great deal of eclipse prediction and mapping software from mainframe (DEC VAX 11/785) to personal computer (Macintosh IIfx) and from one programming language (FORTRAN IV) to another (THINK Pascal). This bulletin is also the first to be printed on a 600 dpi laser printer. The contrast, resolution and readability is a noticeable improvement. The authors have also adopted the international convention of presenting date and time in descending order (i.e. *year, month, day, hour, minute, second*). Word processing and page layout for the publication were done using Microsoft Word v5.1. Figure annotation was done with Claris MacDraw Pro 1.5. Meteorological diagrams were prepared using Windows Draw 3.0 and converted to Macintosh compatible files.

We would like to acknowledge the valued contrbutions of a number of individuals that were essential to the success of this publication. The format and content of the NASA eclipse bulletins has drawn heavily upon over 40 years of eclipse *Circulars* published by the U. S. Naval Observatory. We owe a debt of gratitude to past and present staff of that institution who have performed this service for so many years. In particular, we would like to recognize the work of Julena S. Duncombe, Alan D. Fiala, Marie R. Lukac, John A. Bangert and William T. Harris. The many publications and algorithms of Dr. Jean Meeus have served to inspire a life-long interest in eclipse prediction. We thank Francis Reddy, who helped develop the data base of geographic coordinates for major cities used in the local circumstances predictions. Dr. Wayne Warren graciously provided a draft copy of the *IOTA Observer's Manual* for use in describing contact timings near the path limits. Dr. Jay M. Pasachoff kindly reviewed the manuscript and offered a number of valuable suggestions. The availability of the eclipse bulletins via the Internet is due entirely to the efforts of Dr. Joeseph B. Gurman. The support of Environment Canada is acknowledged in the acquisition and arrangement of the weather data. Finally, the authors thank Goddard's Laboratory for Extraterrestrial Physics for several minutes of CPU time on the LEPVX2 computer. The time was used for verifying predictions generated with the Macintosh.

Permission is freely granted to reproduce any portion of this Reference Publication, including data, figures, maps, tables and text (except for material noted as having been published elsewhere, or by persons other than the authors). All uses and/or publication of this material should be accompanied by an appropriate acknowledgment of the source (e.g. - "Reprinted from *Total Solar Eclipse of 1995 October 24*, Espenak and Anderson, 1994"). The names and spellings of countries, cities and other geopolitical regions are not authoritative, nor do they imply any official recognition in status. Corrections to names, geographic coordinates and elevations are actively solicited in order to update the data base for future eclipses. All calculations, diagrams and opinions presented in this publication are those of the authors and they assume full responsibility for their accuracy.

Fred Espenak	Jay Anderson
NASA/Goddard Space Flight Center	Environment Canada
Planetary Systems Branch, Code 693	900-266 Graham Avenue
Greenbelt, MD 20771	Winnipeg, MB,
USA	CANADA R3C 3V4

Fax:	(301) 286-0212	Fax:	(204) 983-0109
Internet:	u32fe@lepvax.gsfc.nasa.gov	Bitnet:	jander@cc.umanitoba.ca

Current and Future NASA Solar Eclipse Bulletins

NASA Eclipse Bulletin	RP #	Publication Date
Annular Solar Eclipse of 1994 May 10	*1301*	*April 1993*
Total Solar Eclipse of 1994 November 3	*1318*	*October 1993*
Total Solar Eclipse of 1995 October 24	*1344*	*July 1994*
- - - - - - - - - - - future - - - - - - - - - - -		
Total Solar Eclipse of 1997 March 9	—	*Spring 1995*
Total Solar Eclipse of 1998 February 26	—	*Fall-Winter 1995*
Total Solar Eclipse of 1999 August 11	—	*Fall 1996*

TOTAL SOLAR ECLIPSE OF 1995 OCTOBER 24

Table of Contents

Eclipse Predictions...1
    Introduction..1
    Path And Visibility...1
    General Maps of the Eclipse Path..2
        Orthographic Projection Map of the Eclipse Path............................2
        Stereographic Projection Map of the Eclipse Path...........................2
        Equidistant Conic Projection Maps of the Eclipse Path....................3
    Elements, Shadow Contacts and Eclipse Path Tables....................................3
    Local Circumstances Tables...4
    Detailed Maps of the Umbral Path...5
    Estimating Times of Second And Third Contacts..5
    Mean Lunar Radius...6
    Lunar Limb Profile...7
    Limb Corrections To The Path Limits: Graze Zones...................................8
    Saros History..9
Weather Prospects for the Eclipse...10
    Overview...10
    The Middle East..10
    The Indian Subcontinent...10
    Southeast Asia..11
    Weather Summary..12
Observing the Eclipse..13
    Eye Safety During Solar Eclipses..13
    Sky At Totality...14
    Eclipse Photography...14
    Contact Timings from the Path Limits...16
    Plotting the Path on Maps...16
Eclipse Data on Internet..17
    NASA Eclipse Bulletins on Internet...17
    Future Eclipse Paths on Internet...17
Algorithms, Ephemerides and Parameters..18
Bibliography...19
    References..19
    Further Reading..19
Figures..21
Tables...33
Maps of the Umbral Path..63
Request Form for NASA Eclipse Bulletins...73

TOTAL SOLAR ECLIPSE OF 1995 OCTOBER 24

Figures, Tables and Maps

Figures ... 21
- Figure 1: Orthographic Projection Map of the Eclipse Path ... 23
- Figure 2: Stereographic Projection Map of the Eclipse Path ... 24
- Figure 3: The Eclipse Path in Asia ... 25
- Figure 4: The Eclipse Path in India ... 26
- Figure 5: The Eclipse Path in Southeast Asia ... 27
- Figure 6: The Eclipse Path in South China Sea and Celebes Sea ... 28
- Figure 7: The Lunar Limb Profile At 03:30 UT ... 29
- Figure 8: Mean Cloud Cover in October Along the Eclipse Path ... 30
- Figure 9: Frequency of Highly Reflective Clouds in October Along the Path ... 30
- Figure 10: Mean Number of Hours of Sunshine in October for India ... 31
- Figure 11: The Sky During Totality As Seen From Center Line At 03:30 UT ... 32

Tables ... 33
- Table 1: Elements of the Total Solar Eclipse of 1995 October 24 ... 35
- Table 2: Shadow Contacts and Circumstances ... 36
- Table 3: Path of the Umbral Shadow ... 37
- Table 4: Physical Ephemeris of the Umbral Shadow ... 38
- Table 5: Local Circumstances on the Center Line ... 39
- Table 6: Topocentric Data and Path Corrections Due To Lunar Limb Profile ... 40
- Table 7: Mapping Coordinates for the Umbral Path ... 41
- Table 8: Maximum Eclipse and Circumstances for Africa, Middle East and Russia ... 44
- Table 9: Maximum Eclipse and Circumstances for Afghanistan and Pakistan ... 46
- Table 10: Maximum Eclipse and Circumstances for Indian Asia – I ... 48
- Table 11: Maximum Eclipse and Circumstances for Indian Asia – II ... 50
- Table 12: Maximum Eclipse and Circumstances for Southeast Asia ... 52
- Table 13: Maximum Eclipse and Circumstances for the Orient ... 54
- Table 14: Maximum Eclipse and Circumstances for Indonesia, Japan and Malaysia ... 56
- Table 15: Maximum Eclipse and Circumstances for Australia, Philippines and Pacific ... 58
- Table 16: Climatological Statistics Along the Eclipse Path ... 60
- Table 17: Solar Eclipse Exposure Guide ... 61

Maps of the Umbral Path ... 63
- Map 1: Iran and Afghanistan ... 65
- Map 2: Afghanistan, Pakistan and India ... 66
- Map 3: India ... 67
- Map 4: India, Bangladesh and Myanmar ... 68
- Map 5: Myanmar, Thailand and Cambodia ... 69
- Map 6: Cambodia and Viet Nam ... 70
- Map 7: Borneo ... 71
- Map 8: Celebes Sea ... 72

ECLIPSE PREDICTIONS

INTRODUCTION

On Tuesday, 1995 October 24, a total eclipse of the Sun will be visible from much of the eastern hemisphere. The path of the Moon's umbral shadow begins in the Middle East, and sweeps across India, Southeast Asia, and South China Sea where it briefly engulfs the northern tip of Borneo before swinging eastward between the Philippines and New Guinea. The path ends at sunset in the Pacific Ocean south of the Marshall Islands. A partial eclipse will be seen within the much broader path of the Moon's penumbral shadow, which includes most of Asia, the Indonesian archipelago and Australia (Figures 1 and 2).

PATH AND VISIBILITY

The path of the Moon's umbral shadow begins in central Iran about 100 kilometers south of Tehran near the city of Oom (Figure 3). As the shadow first contacts Earth along the sunrise terminator (2:52 UT), the path is a scant 16 kilometers wide and the total eclipse barely lasts 16 seconds. During the first seven minutes of central eclipse, the umbra travels to the southeast and quickly sweeps through Afghanistan and Pakistan where it enters northern India at 3:00 UT (Figure 4). Moving with a surface velocity of 2.255 km/s, the shadow's path width has increased to 37 kilometers while the duration of totality has nearly tripled to 44 seconds. The early morning Sun then stands 21° above the eastern horizon. Four minutes later, a partial eclipse maximum of magnitude 0.965 will be seen from New Dehli, which lies 130 kilometers north of the path. The umbra's northern edge narrowly misses the city of Agra, home of one of the world's greatest architectural achievements, the renowned Taj Mahal. From the grounds of this remarkable structure, a partial eclipse with a tantalizing magnitude of 0.997 peaks at 3:05 UT.

Allahabad lies near the path's northern limit where mid-totality occurs at 3:09 UT and lasts a brief 31 seconds. The center line is only twenty kilometers south, where totality is 1 minute 3 seconds. By the time the shadow reaches the Ganges delta (3:20 UT), the path is 55 kilometers wide, the duration of the total eclipse is 1 minute 18 seconds and the Sun has climbed to an altitude of 40°. Calcutta straddles the path's northern limit, where 11 million people could witness a spectacular grazing event.

The umbra leaves India and sweeps across the Bay of Bengal where it reaches the western coast of Myanmar at 3:30 UT (Figure 5). The path's width increases to 61 kilometers and the duration is 1 minute 30 seconds. As it proceeds inland, the shadow passes 100 kilometers north of Yangon, which experiences a 0.975 magnitude eclipse at 3:38 UT. Traveling southeast, the umbra crosses the Thaungyin River and enters Thailand at 3:44 UT. The duration on the center line is then 1 minute 45 seconds, the Sun stands 54° and the umbra travels with a speed of 0.73 km/s. The Thai capital of Bangkok lies 140 kilometers to the south of the path and experiences a partial eclipse of magnitude 0.957 at 3:51 UT. The path width and central duration continue to increase modestly as the shadow sweeps across central Cambodia. At 4:00 UT, the center line duration is 1 minute 57 seconds, the path width is 72 kilometers and the Sun's altitude is 62°. The ancient ruins of Angkor Wat lie deep in the path and will bear mute witness to the celestial spectacle. Leaving Cambodia at 4:09 UT, the umbra sweeps across southern Viet Nam and passes 100 kilometers north of Ho Chi Minh City, where a partial eclipse of magnitude 0.978 occurs.

Reaching the southeastern coastline of Viet Nam at 4:14 UT, the shadow leaves the mainland and heads out across the South China Sea (Figure 6). The instant of greatest eclipse[1] occurs shortly thereafter at 4:32:29.5 UT. At that point, the length of totality reaches its maximum duration of 2 minutes 10 seconds, the Sun's altitude is 69°, the path width is 78 kilometers and the umbra's velocity is 0.564 km/s. The next landfall occurs in Sarawak along the northern coast of Borneo at 04:44 UT. Following the northern coastline, the shadow soon leaves land again. Its southern limit bisects the island of Tawi-Tawi as it crosses the Celebes Sea. The island of Pulau Sangihe is the last significant landfall as the shadow passes 150 kilometers south of Mindanao at approximately 5:15 UT.

The path continues east across the Pacific as its width and duration dwindle. The umbra leaves Earth at sunset at 6:13 UT, where the total phase lasts 20 seconds. Due to the particularly narrow width of this path, less than 0.1% of Earth's surface area falls within the 13 600 kilometer long eclipse track.

---

[1] The instant of greatest eclipse occurs when the distance between the Moon's shadow axis and Earth's geocenter reaches a minimum. Although greatest eclipse differs slightly from the instants of greatest magnitude and greatest duration (for total eclipses), the differences are usually negligible.

GENERAL MAPS OF THE ECLIPSE PATH

ORTHOGRAPHIC PROJECTION MAP OF THE ECLIPSE PATH

Figure 1 is an orthographic projection map of Earth [adapted from Espenak, 1987] showing the path of penumbral (partial) and umbral (total) eclipse. The daylight terminator is plotted for the instant of greatest eclipse with north at the top. The sub-Earth point is centered over the point of greatest eclipse and is marked with an asterisk at **GE**. Earth's sub-solar point at that instant is also indicated by the point **SS**.

The limits of the Moon's penumbral shadow define the region of visibility of the partial eclipse. This saddle shaped region often covers more than half of Earth's daylight hemisphere and consists of several distinct zones or limits. At the northern and/or southern boundaries lie the limits of the penumbra's path. Partial eclipses have only one of these limits, as do central eclipses when the shadow axis falls no closer than about 0.45 radii from Earth's center. Great loops at the western and eastern extremes of the penumbra's path identify the areas where the eclipse begins/ends at sunrise and sunset, respectively. If the penumbra has both a northern and southern limit, the rising and setting curves form two separate, closed loops. Otherwise, the curves are connected in a distorted figure eight. Bisecting the 'eclipse begins/ends at sunrise and sunset' loops is the curve of maximum eclipse at sunrise (western loop) and sunset (eastern loop). The exterior tangency points **P1** and **P4** mark the coordinates where the penumbral shadow first contacts (partial eclipse begins) and last contacts (partial eclipse ends) Earth's surface. If the penumbral path has both a northern and southern limit (as does the October 1995 eclipse), then the interior tangency points **P2** and **P3** are also plotted and correspond to the coordinates where the penumbral cone becomes internally tangent to Earth's disk. Likewise, the points **U1** and **U2** mark the exterior and interior coordinates where the umbral shadow initially contacts Earth (path of total eclipse begins). The points **U3** and **U4** mark the interior and exterior points of final umbral contact with Earth's surface (path of total eclipse ends).

A curve of maximum eclipse is the locus of all points where the eclipse is at maximum at a given time. They are plotted at each half hour Universal Time (UT), and generally run from northern to southern penumbral limits, or from the maximum eclipse at sunrise or sunset curves to one of the limits. The outline of the umbral shadow is plotted every ten minutes in UT. Curves of constant eclipse magnitude[2] delineate the locus of all points where the magnitude at maximum eclipse is constant. These curves run exclusively between the curves of maximum eclipse at sunrise and sunset. Furthermore, they are parallel to the northern/southern penumbral limits and the umbral paths of central eclipses. Northern and southern limits of the penumbra may be thought of as curves of constant magnitude of 0%, while adjacent curves are for magnitudes of 20%, 40%, 60% and 80%. The northern and southern limits of the path of total eclipse are curves of constant magnitude of 100%.

At the top of Figure 1, the Universal Time of geocentric conjunction between the Moon and Sun is given followed by the instant of greatest eclipse. The eclipse magnitude is given for greatest eclipse. For central eclipses (both total and annular), it is equivalent to the geocentric ratio of diameters of the Moon and Sun. Gamma is the minimum distance of the Moon's shadow axis from Earth's center in units of equatorial Earth radii. The shadow axis passes south of Earth's geocenter for negative values of Gamma. Finally, the Saros series number of the eclipse is given along with its relative sequence in the series.

STEREOGRAPHIC PROJECTION MAP OF THE ECLIPSE PATH

The stereographic projection of Earth in Figure 2 depicts the path of penumbral and umbral eclipse in greater detail. The map is oriented with the point of greatest eclipse near the center and north is at the top. International political borders are shown and circles of latitude and longitude are plotted at 20° increments. The region of penumbral or partial eclipse is identified by its northern and southern limits, curves of eclipse begins or ends at sunrise and sunset, and curves of maximum eclipse at sunrise and sunset. Curves of constant eclipse magnitude are plotted for 20%, 40%, 60% and 80%, as are the limits of the path of total eclipse. Also included are curves of greatest eclipse at every half hour Universal Time.

Figures 1 and 2 may be used to quickly determine the approximate time and magnitude of maximum eclipse at any location within the eclipse path.

---

[2] Eclipse magnitude is defined as the fraction of the Sun's diameter occulted by the Moon. It is usually expressed at greatest eclipse. Eclipse magnitude is strictly a ratio of *diameters* and should not be confused with eclipse obscuration which is a measure of the Sun's surface *area* occulted by the Moon. Eclipse magnitude may be expressed as either a percentage or a decimal fraction (e.g.: 50% or 0.5).

EQUIDISTANT CONIC PROJECTION MAPS OF THE ECLIPSE PATH

Figures 3, 4, 5 and 6 are equidistant conic projection maps that isolate specific regions of the eclipse path. The projection was selected to minimize distortion over the regions depicted. Once again, curves of maximum eclipse and constant eclipse magnitude are plotted and labeled. A linear scale is included for estimating approximate distances (kilometers) in each figure. Within the northern and southern limits of the path of totality, the outline of the umbral shadow is plotted at five minute intervals. The center line duration of totality appears near the umbra at various points along the path.

Figure 3 is drawn at a scale of ~1:11,060,000, while Figures 4, 5 and 6 are drawn at a scale of ~1:3,250,000. All four figures include the positions of many of the larger cities or metropolitan areas in and near the central path. The size of each city is logarithmically proportional to its population according to 1990 census data (Rand McNally, 1991).

ELEMENTS, SHADOW CONTACTS AND ECLIPSE PATH TABLES

The geocentric ephemeris for the Moon and Sun, various parameters and constants used in the predictions, and the Besselian elements (polynomial form) are given in Table 1. The eclipse elements and predictions were derived from the DE200 and LE200 ephemerides (solar and lunar, respectively) developed jointly by the Jet Propulsion Laboratory and the U. S. Naval Observatory for use in the *Astronomical Almanac* for 1984 and thereafter. Unless otherwise stated, all predictions are based on center of mass positions for the Moon and Sun with no corrections made for center of figure, lunar limb profile or atmospheric refraction. The predictions depart from normal IAU convention through the use of a smaller constant for the mean lunar radius k for all umbral contacts (see: LUNAR LIMB PROFILE). Times are expressed in either Terrestrial Dynamical Time (TDT) or in Universal Time (UT), where the best value of ΔT[3] available at the time of preparation is used.

Table 2 lists all external and internal contacts of penumbral and umbral shadows with Earth. They include TDT times and geodetic coordinates with and without corrections for ΔT. The contacts are defined:

P1 - Instant of first external tangency of penumbral shadow cone with Earth's limb.
(partial eclipse begins)
P2 - Instant of first internal tangency of penumbral shadow cone with Earth's limb.
P2 - Instant of last internal tangency of penumbral shadow cone with Earth's limb.
P4 - Instant of last external tangency of penumbral shadow cone with Earth's limb.
(partial eclipse ends)

U1 - Instant of first external tangency of umbral shadow cone with Earth's limb.
(umbral eclipse begins)
U2 - Instant of first internal tangency of umbral shadow cone with Earth's limb.
U2 - Instant of last internal tangency of umbral shadow cone with Earth's limb.
U4 - Instant of last external tangency of umbral shadow cone with Earth's limb.
(umbral eclipse ends)

Similarly, the northern and southern extremes of the penumbral and umbral paths, and extreme limits of the umbral center line are given. The IAU longitude convention is used throughout this publication (i.e. - eastern longitudes are positive; western longitudes are negative; negative latitudes are south of the Equator).

The path of the umbral shadow is delineated at five minute intervals in Universal Time in Table 3. Coordinates of the northern limit, the southern limit and the center line are listed to the nearest tenth of an arc-minute (~185 m at the Equator). The path azimuth, path width and umbral duration are calculated for the center line position. The path azimuth is the direction of the umbral shadow's motion projected onto the surface of the Earth. Table 4 presents a physical ephemeris for the umbral shadow at five minute intervals in UT. The center line coordinates are followed by the topocentric ratio of the apparent diameters of the Moon and Sun, the eclipse obscuration[4], and the Sun's altitude and azimuth at that instant. The central path width, the umbral shadow's major and minor axes and its instantaneous velocity with respect to Earth's surface are included. Finally, the center line duration of the umbral phase is given.

---

[3] ΔT is the difference between Terrestrial Dynamical Time and Universal Time
[4] Eclipse obscuration is defined as the fraction of the Sun's surface area occulted by the Moon.

Local circumstances for each center line position listed in Tables 3 and 4 are presented in Table 5. The first three columns give the Universal Time of maximum eclipse, the center line duration of totality and the altitude of the Sun at that instant. The following columns list each of the four eclipse contact times followed by their related contact position angles and the corresponding altitude of the Sun. The four contacts identify significant stages in the progress of the eclipse. They are defined as follows:

> **First Contact**   –   Instant of first external tangency between the Moon and Sun.
> (partial eclipse begins)
>
> **Second Contact**   –   Instant of first internal tangency between the Moon and Sun.
> (central or umbral eclipse begins; total or annular eclipse begins)
>
> **Third Contact**   –   Instant of last internal tangency between the Moon and Sun.
> (central or umbral eclipse ends; total or annular eclipse ends)
>
> **Fourth Contact**   –   Instant of last external tangency between the Moon and Sun.
> (partial eclipse ends)

The position angles **P** and **V** identify the point along the Sun's disk where each contact occurs[5]. The altitude of the Sun at second and third contact is omitted since it is always within 1° of the altitude at maximum eclipse (column 3).

Table 6 presents topocentric values from the central path at maximum eclipse for the Moon's horizontal parallax, semi-diameter, relative angular velocity with respect to the Sun, and libration in longitude. The altitude and azimuth of the Sun are given along with the azimuth of the umbral path. The northern limit position angle identifies the point on the lunar disk defining the umbral path's northern limit. It is measured counter-clockwise from the north point of the Moon. In addition, corrections to the path limits due to the lunar limb profile are listed. The irregular profile of the Moon results in a zone of 'grazing eclipse' at each limit that is delineated by interior and exterior contacts of lunar features with the Sun's limb. This geometry is described in greater detail in the section LIMB CORRECTIONS TO THE PATH LIMITS: GRAZE ZONES. Corrections to center line durations due to the lunar limb profile are also included. When added to the durations in Tables 3, 4, 5 and 7, a slightly shorter central total phase is predicted.

To aid and assist in the plotting of the umbral path on large scale maps, the path coordinates are also tabulated at 1° intervals in longitude in Table 7. The latitude of the northern limit, southern limit and center line for each longitude is tabulated along with the Universal Time of maximum eclipse at each position. Finally, local circumstances on the center line at maximum eclipse are listed and include the Sun's altitude and azimuth, the umbral path width and the central duration of totality.

LOCAL CIRCUMSTANCES TABLES

Local circumstances from approximately 400 cities, metropolitan areas and places in Africa, Asia, the Indonesian archipelago and Australia are presented in Tables 8 through 15. Each table is broken down into two parts. The first part, labeled **a**, appears on even numbered pages and gives circumstances at maximum eclipse[6] for each location. The coordinates are listed along with the location's elevation (meters) above sea-level, if known. If the elevation is unknown (i.e. - not in the data base), then the local circumstances for that location are calculated at sea-level. In any case, the elevation does not play a significant role in the predictions unless the location is near the umbral path limits and the Sun's altitude is relatively small (<15°). The Universal Time of maximum eclipse (either partial or total) is listed to an accuracy of 0.1 seconds. If the eclipse is total, then the umbral duration and the path width are given. Next, the altitude and azimuth of the Sun at maximum eclipse are listed along with the position angles **P** and **V** of the Moon's disk with respect to the Sun. Finally, the magnitude and obscuration are listed at the instant of maximum eclipse. Note that for umbral eclipses (annular and total), the eclipse magnitude is identical to the topocentric ratio of the Moon's and Sun's apparent diameters. Furthermore, the eclipse magnitude is always less than 1 for annular eclipses and equal to or greater than 1 for total eclipses.

The second part of each table, labeled **b**, is found on odd numbered pages. It gives local circumstances at each eclipse contact for every location listed in part **a**. The Universal Time of each contact

---

[5] P is defined as the contact angle measured counter-clockwise from the *north* point of the Sun's disk.

V is defined as the contact angle measured counter-clockwise from the *zenith* point of the Sun's disk.

[6] For partial eclipses, maximum eclipse is the instant when the greatest fraction of the Sun's diameter is occulted. For umbral eclipses (total or annular), maximum eclipse is the instant of mid-totality or mid-annularity.

is given along with the altitude of the Sun, followed by position angles **P** and **V**. These angles identify the point along the Sun's disk where each contact occurs and are measured counter-clockwise from the north and zenith points, respectively. Locations outside the umbral path miss the umbral eclipse and only witness first and fourth contacts. The effects of refraction have not been included in these calculations, nor have there been corrections for center of figure or the lunar limb profile.

Since the track of this eclipse is especially narrow (<80 km), few cities actually fall within the path. Locations were chosen based on general geographic distribution, population, and proximity near or within the central path. The primary source for geographic coordinates is *The New International Atlas* (Rand McNally, 1991). Elevations for major cities were taken from *Climates of the World* (U. S. Dept. of Commerce, 1972). In this rapidly changing political world, it is often difficult to ascertain the correct name or spelling for a given location. Therefore, the information presented here is for location purposes only and is not meant to be authoritative. Furthermore, it does not imply recognition of status of any location by the United States Government. Corrections to names, spellings, coordinates and elevations is solicited in order to update the geographic data base for future eclipse predictions.

DETAILED MAPS OF THE UMBRAL PATH

The path of totality has been plotted by hand on a set of eight detailed maps appearing in the last section of this publication. The maps are Global Navigation and Planning Charts or GNC's from the Defense Mapping Agency, which use a Lambert conformal conic projection. More specifically, GNC-12 covers the Middle East and Indian section of the path while GNC-13 covers Southeast Asia and Indonesia. GNC's have a scale of 1:5,000,000 (1 inch ~ 69 nautical miles), which is adequate for showing major cities, highways, airports, rivers, bodies of water and basic topography required for eclipse expedition planning including site selection, transportation logistics and weather contingency strategies.

Northern and southern limits as well as the center line of the path are plotted using Table 7. Although no corrections have been made for center of figure or lunar limb profile, they have little or no effect at this scale. Atmospheric refraction has not been included as its effects play a significant role only at low solar altitudes (<15°). In any case, refraction corrections to the path are uncertain since they depend on the atmospheric temperature-pressure profile, which cannot be predicted in advance. If observations from the graze zones are planned, then the path must be plotted on higher scale maps using limb corrections in Table 6. See PLOTTING THE PATH ON MAPS for sources and more information. The GNC paths also depict the curve of maximum eclipse at five minute increments in Universal Time from Table 3.

ESTIMATING TIMES OF SECOND AND THIRD CONTACTS

The times of second and third contact for any location not listed in this publication can be estimated using the detailed maps found in the final section. Alternatively, the contact times can be estimated from maps on which the umbral path has been plotted. Table 7 lists the path coordinates conveniently arranged in 1° increments of longitude to assist plotting by hand. The path coordinates in Table 3 define a line of maximum eclipse at five minute increments in time. These lines of maximum eclipse each represent the projection diameter of the umbral shadow at the given time. Thus, any point on one of these lines will witness maximum eclipse (i.e.: mid-totality) at the same instant. The coordinates in Table 3 should be added to the map in order to construct lines of maximum eclipse.

The estimation of contact times for any one point begins with an interpolation for the time of maximum eclipse at that location. The time of maximum eclipse is proportional to a point's distance between two adjacent lines of maximum eclipse, measured along a line parallel to the center line. This relationship is valid along most of the path with the exception of the extreme ends, where the shadow experiences its largest acceleration. The center line duration of totality **D** and the path width **W** are similarly interpolated from the values of the adjacent lines of maximum eclipse as listed in Table 3. Since the location of interest probably does not lie on the center line, it is useful to have an expression for calculating the duration of totality **d** as a function of its perpendicular distance **a** from the center line:

$$\mathbf{d} = \mathbf{D} \cdot (1 - (2\,\mathbf{a}/\mathbf{W})^2)^{1/2} \text{ seconds} \qquad [1]$$

where: **d** = duration of totality at desired location (seconds)
   D = duration of totality on the center line (seconds)
   a = perpendicular distance from the center line (kilometers)
   W = width of the path (kilometers)

If t_m is the interpolated time of maximum eclipse for the location, then the approximate times of second and third contacts (t_2 and t_3, respectively) are:

Second Contact: $\qquad t_2 = t_m - d/2 \qquad$ [2]
Third Contact: $\qquad t_3 = t_m + d/2 \qquad$ [3]

The position angles of second and third contact (either **P** or **V**) for any location off the center line are also useful in some applications. First, linearly interpolate the center line position angles of second and third contacts from the values of the adjacent lines of maximum eclipse as listed in Table 5. If X_2 and X_3 are the interpolated center line position angles of second and third contacts, then the position angles x_2 and x_3 of those contacts for an observer located **a** kilometers from the center line are:

Second Contact: $\qquad x_2 = X_2 - \text{ArcSin}(2\,a/W) \qquad$ [4]
Third Contact: $\qquad x_3 = X_3 + \text{ArcSin}(2\,a/W) \qquad$ [5]

where: x_n = the interpolated position angle (either **P** or **V**) of contact **n** at location
X_n = the interpolated position angle (either **P** or **V**) of contact **n** on center line
a = perpendicular distance from the center line (kilometers)
 (use negative values for locations south of the center line)
W = width of the path (kilometers)

MEAN LUNAR RADIUS

A fundamental parameter used in the prediction of solar eclipses is the Moon's mean radius **k**, expressed in units of Earth's equatorial radius. The actual radius of the Moon varies as a function of position angle and libration due to the irregularity of the lunar limb profile. From 1968 through 1980, the Nautical Almanac Office used two separate values for **k** in their eclipse predictions. The larger value (**k**=0.2724880), representing a mean over lunar topographic features, was used for all penumbral (i.e. - exterior) contacts and for annular eclipses. A smaller value (**k**=0.272281), representing a mean minimum radius, was reserved exclusively for umbral (i.e. - interior) contact calculations of total eclipses [*Explanatory Supplement*, 1974]. Unfortunately, the use of two different values of **k** for umbral eclipses introduces a discontinuity in the case of hybrid or annular-total eclipses.

In August 1982, the IAU General Assembly adopted a value of **k**=0.2725076 for the mean lunar radius. This value is currently used by the Nautical Almanac Office for all solar eclipse predictions [Fiala and Lukac, 1983] and is currently the best mean radius, averaging mountain peaks and low valleys along the Moon's rugged limb. In general, the adoption of one single value for **k** is commendable because it eliminates the discontinuity in the case of annular-total eclipses and ends confusion arising from the use of two different values. However, the use of even the best 'mean' value for the Moon's radius introduces a problem in predicting the true character and duration of umbral eclipses, particularly total eclipses. A total eclipse can be defined as an eclipse in which the Sun's disk is completely occulted by the Moon. This cannot occur so long as any photospheric rays are visible through deep valleys along the Moon's limb [Meeus, Grosjean and Vanderleen, 1966]. But the use of the IAU's mean **k** guarantees that some annular or annular-total eclipses will be misidentified as total. A case in point is the eclipse of 3 October 1986. The *Astronomical Almanac* identified this event as a total eclipse of 3 seconds duration when it was, in fact, a beaded annular eclipse. Clearly, a smaller value of **k** is needed since it is more representative of the deeper lunar valley floors, hence the minimum solid disk radius and helps ensure that an eclipse is truly total.

Of primary interest to most observers are the times when central eclipse begins and ends (second and third contacts, respectively) and the duration of the central phase. When the IAU's mean value for **k** is used to calculate these times, they must be corrected to accommodate low valleys (total) or high mountains (annular) along the Moon's limb. The calculation of these corrections is not trivial but must be performed, especially if one plans to observe near the path limits [Herald, 1983]. For observers near the center line of a total eclipse, the limb corrections can be more closely approximated by using a smaller value of **k** which accounts for the valleys along the profile.

This publication uses the IAU's accepted value of **k** (**k**=0.2725076) for all penumbral (exterior) contacts. In order to avoid eclipse type misidentification and to predict central durations which are closer to the actual durations for at total eclipses, we depart from convention by adopting the smaller value for **k** (**k**=0.272281) for all umbral (interior) contacts. This is consistent with predictions in *Fifty Year Canon of Solar Eclipses: 1986 - 2035* [Espenak, 1987]. Consequently, the smaller **k** produces shorter umbral durations and narrower paths for total eclipses when compared with calculations using the IAU **k** value. Similarly, the smaller **k** predicts longer umbral durations and wider paths for annular eclipses than does the IAU **k**.

LUNAR LIMB PROFILE

Eclipse contact times, the magnitude and the duration of totality all ultimately depend on the angular diameters and relative velocities of the Moon and Sun. Unfortunately, these calculations are limited in accuracy by the departure of the Moon's limb from a perfectly circular figure. The Moon's surface exhibits a rather dramatic topography, that manifests itself as an irregular limb when seen in profile. Most eclipse calculations assume some mean lunar radius that averages high mountain peaks and low valleys along the Moon's rugged limb. Such an approximation is acceptable for many applications, but if higher accuracy is needed, the Moon's actual limb profile must be considered. Fortunately, an extensive body of knowledge exists on this subject in the form of Watts' limb charts [Watts, 1963]. These data are the product of a photographic survey of the marginal zone of the Moon and give limb profile heights with respect to an adopted smooth reference surface (or datum). Analyses of lunar occultations of stars by Van Flandern [1970] and Morrison [1979] have shown that the average cross-section of Watts' datum is slightly elliptical rather than circular. Furthermore, the implicit center of the datum (i.e. - the center of figure) is displaced from the Moon's center of mass. In a follow-up analysis of 66000 occultations, Morrison and Appleby [1981] have found that the radius of the datum appears to vary with libration. These variations produce systematic errors in Watts' original limb profile heights that attain 0.4 arc-seconds at some position angles. Thus, corrections to Watts' limb profile data are necessary to ensure that the reference datum is a sphere with its center at the center of mass.

The Watts charts have been digitized by Her Majesty's Nautical Almanac Office in Herstmonceux, England, and transformed to grid-profile format at the U. S. Naval Observatory. In this computer readable form, the Watts limb charts lend themselves to the generation of limb profiles for any lunar libration. Ellipticity and libration corrections may be applied to refer the profile to the Moon's center of mass. Such a profile can then be used to correct eclipse predictions which have been generated using a mean lunar limb.

Along the 1995 eclipse path, the Moon's topocentric libration (physical + optical libration) in longitude ranges from $l=-3.2°$ to $l=-4.9°$. Thus, a limb profile with the appropriate libration is required in any detailed analysis of contact times, central durations, etc.. Nevertheless, a profile with an intermediate libration is valuable for general planning purposes. The lunar limb profile presented in Figure 7 includes corrections for center of mass and elipticity [Morrison and Appleby, 1981]. It is generated for is for 3:30 UT, which corresponds to the western coast of Myanmar. The Moon's topocentric libration in longitude is $l=-3.52°$, and the topocentric semi-diameters of the Sun and Moon are 964.7 and 981.4 arc-seconds, respectively. The Moon's angular velocity with respect to the Sun is 0.370 arc-seconds per second.

The radial scale of the limb profile in Figure 7 (at bottom) is greatly exaggerated so that the true limb's departure from the mean lunar limb is readily apparent. The mean limb with respect to the center of figure of Watts' original data is shown (dashed) along with the mean limb with respect to the center of mass (solid). Note that all the predictions presented in this publication are calculated with respect to the latter limb unless otherwise noted. Position angles of various lunar features can be read using the protractor in the center of the diagram. The position angles of second and third contact are clearly marked along with the north pole of the Moon's axis of rotation and the observer's zenith at mid-totality. The dashed line arrows identify the points on the limb which define the northern and southern limits of the path. To the upper left of the profile are the Sun's topocentric coordinates at maximum eclipse. They include the right ascension **R.A.**, declination **Dec.**, semi-diameter **S.D.** and horizontal parallax **H.P.**. The corresponding topocentric coordinates for the Moon are to the upper right. Below and left of the profile are the geographic coordinates of the center line at 3:30 UT while the times of the four eclipse contacts at that location appear to the lower right. Directly below the profile are the local circumstances at maximum eclipse. They include the Sun's altitude and azimuth, the path width, and central duration. The position angle of the path's northern/southern limit axis is **PA(N.Limit)** and the angular velocity of the Moon with respect to the Sun is **A.Vel.(M:S)**. At the bottom left are a number of parameters used in the predictions, and the topocentric lunar librations appear at the lower right.

In investigations where accurate contact times are needed, the lunar limb profile can be used to correct the nominal or mean limb predictions. For any given position angle, there will be a high mountain (annular eclipses) or a low valley (total eclipses) in the vicinity that ultimately determines the true instant of contact. The difference, in time, between the Sun's position when tangent to the contact point on the mean limb and tangent to the highest mountain (annular) or lowest valley (total) at actual contact is the desired correction to the predicted contact time. On the exaggerated radial scale of Figure 7, the Sun's limb can be represented as an epicyclic curve that is tangent to the mean lunar limb at the point of contact and departs from the limb by **h** as follows:

$$h = S (m-1) (1-\cos[C]) \qquad [6]$$

where: h = departure of Sun's limb from mean lunar limb
S = Sun's semi-diameter
m = eclipse magnitude
C = angle from the point of contact

Herald [1983] has taken advantage of this geometry to develop a graphical procedure for estimating correction times over a range of position angles. Briefly, a displacement curve of the Sun's limb is constructed on a transparent overlay by way of equation [6]. For a given position angle, the solar limb overlay is moved radially from the mean lunar limb contact point until it is tangent to the lowest lunar profile feature in the vicinity. The solar limb's distance **d** (arc-seconds) from the mean lunar limb is then converted to a time correction Δ by:

$$\Delta = d\, v\, \cos[X - C] \qquad [7]$$

where: Δ = correction to contact time (seconds)
d = distance of Solar limb from Moon's mean limb (arc-sec)
v = angular velocity of the Moon with respect to the Sun (arc-sec/sec)
X = center line position angle of the contact
C = angle from the point of contact

This operation may be used for predicting the formation and location of Baily's beads. When calculations are performed over a large range of position angles, a contact time correction curve can then be constructed.

Since the limb profile data are available in digital form, an analytic solution to the problem is possible that is straightforward and quite robust. Curves of corrections to the times of second and third contact for most position angles have been computer generated and are plotted in Figure 7. In interpreting these curves, the circumference of the central protractor functions as the nominal or mean contact time (i.e.: calculated using the Moon's mean limb) as a function of position angle. The departure of the correction curve from the mean contact time can then be read directly from Figure 7 for any position angle by using the radial scale at bottom right (units in seconds of time). Time corrections external to the protractor (about half of all second contact corrections) are added to the mean contact time; time corrections internal to the protractor (all third contact corrections) are subtracted from the mean contact time.

Throughout most of Asia, the Moon's topocentric libration in longitude at maximum eclipse is within 0.3° of its value at 3:30 UT. Therefore, the limb profile and contact correction time curves in Figure 7 may be used in all but the most critical investigations.

LIMB CORRECTIONS TO THE PATH LIMITS: GRAZE ZONES

The northern and southern umbral limits provided in this publication were derived using the Moon's center of mass and a mean lunar radius. They have not been corrected for the Moon's center of figure or the effects of the lunar limb profile. In applications where precise limits are required, Watts' limb data must be used to correct the nominal or mean path. Unfortunately, a single correction at each limit is not possible since the Moon's libration in longitude and the contact points of the limits along the Moon's limb each vary as a function of time and position along the umbral path. This makes it necessary to calculate a unique correction to the limits at each point along the path. Furthermore, the northern and southern limits of the umbral path are actually paralleled by a relatively narrow zone where the eclipse is neither penumbral nor umbral. An observer positioned here will witness a slender solar crescent that is fragmented into a series of bright beads and short segments whose morphology changes quickly with the rapidly varying geometry of the Moon with respect to the Sun. These beading phenomena are caused by the appearance of photospheric rays that alternately pass through deep lunar valleys and hide behind high mountain peaks as the Moon's irregular limb grazes the edge of the Sun's disk. The geometry is directly analogous to the case of grazing occultations of stars by the Moon. The graze zone is typically five to ten kilometers wide and its interior and exterior boundaries can be predicted using the lunar limb profile. The interior boundaries define the actual limits of the umbral eclipse (both total and annular) while the exterior boundaries set the outer limits of the grazing eclipse zone.

Table 6 provides topocentric data and corrections to the path limits due to the true lunar limb profile. At five minute intervals, the table lists the Moon's topocentric horizontal parallax, semi-diameter, relative angular velocity of the Moon with respect to the Sun and lunar libration in longitude. The Sun's center line altitude and azimuth is given, followed by the azimuth of the umbral path. The position angle of the point on the Moon's limb which defines the northern limit of the path is measured counter-clockwise (i.e. - eastward) from the north point on the limb. The path corrections to the northern and southern limits are listed as interior and exterior components in order to define the graze zone. Positive corrections are in the northern sense while negative shifts are in the southern sense. These corrections (minutes of arc in latitude) may be added directly to the path coordinates listed in Table 3. Corrections to the center line umbral durations due to the lunar limb profile are also included and they are all negative. Thus, when added to the central durations given in Tables 3, 4, 5 and 7, a slightly shorter central total phase is predicted.

SAROS HISTORY

The total eclipse of 1995 October 24 is the twenty-second member of Saros series 143, as defined by van den Bergh [1955]. All eclipses in the series occur at the Moon's ascending node and gamma[7] decreases with each member in the series. The family began on 1617 Mar 2 with a partial eclipse at high latitudes in the northern hemisphere. During the first one and a half centuries, ten partial eclipses occurred with the eclipse magnitude of each succeeding event gradually increasing. Finally, the first umbral eclipse occurred on 1797 Jun 24. The event was a total eclipse visible from the Arctic Ocean and eastern Siberia. Perhaps the most remarkable characteristic of this eclipse was its unusual umbral path nearly 1000 kilometers wide. During the 1800's, the series continued producing total eclipses whose maximum durations gradually increasing to nearly four minutes.

Some of the eclipses of Saros 143 have contributed significantly to our understanding of the Sun. For instance, during the total eclipse of 1851 July 28, Airy described the Sun's chromosphere in detail, while Grant, Swan and von Littrow determined that prominences were a physical part of the Sun, rather than the Moon. The first photograph of the corona, a daguerreotype, was made in Prussia at this event. The following eclipse of 1869 Aug 7 passed centrally through the United States and is notable for the major scientific expeditions organized to study it. Young and Harkness independently discovered a mysterious, bright green line in the corona's spectrum. It wasn't until 1941 that Edlén identified the line as iron that has lost 13 electrons (Fe XIV). In Russia, Mendeleev used a balloon to ascend above the clouds to observe the total eclipse of 1887 Aug 19. One saros period later, the umbra's path crossed through Spain during the well observed eclipse of 1905 Aug 30.

Although each succeeding path was shifting south towards the equator, the duration of totality began dropping. This was due to the Moon's progressively increasing distance from Earth as each eclipse occurred nearer to apogee. By 1977 Oct 12, the duration had dropped below three minutes. The 1995 Oct 24 event is the twelfth and last total eclipse of Saros 143. The next event of 2013 Nov 3 is a hybrid eclipse since it is total along most of its path but becomes annular near the sunrise and sunset portions of the track. The following three events are each annular/total as the path of totality grows progressively narrower and shorter. Finally, the series produces its first entirely annular eclipse on 2085 Dec 16.

During the next two and a half centuries, the duration of annularity gradually increases as the paths regress northward. The trend north is due to the passage of Earth through the vernal equinox which shifts the northern hemisphere southward with respect to the geocenter. The paths resume their southern migration with the eclipse of 2338 May 20. The duration of annularity now exceeds two minutes. The remaining eleven annular members of the series possess paths that shift progressively south while the duration gradually rises above four minutes. The final annular eclipse occurs on 2536 Sep 16 with a duration of 4 minutes 48 seconds. As the series winds down, it produces twenty more partial eclipses at high southern latitudes. Saros 143 finally ends with its seventy-second event on 2897 Apr 23.

In summary, Saros series 143 includes 72 eclipses with the following distribution:

Saros 143	Partial	Annular	Ann/Total	Total
Non-Central	30	0	0	0
Central	—	26	4	12

---

[7] Gamma is measured in Earth radii and is the minimum distance of the Moon's shadow axis from Earth's center during an eclipse. This occurs at and defines the instant of greatest eclipse. Gamma takes on negative values when the shadow axis is south of the Earth's center.

WEATHER PROSPECTS FOR THE ECLIPSE

OVERVIEW

This eclipse begins at moderate latitudes over Iran, heading steadily southeastward toward the equator for most of its length. At its beginning, weather patterns are influenced by high and low pressure systems moving in the upper westerly flow, just as in North America and Europe. Through India and Southeast Asia, the upper westerlies lose their importance, and the track moves into a region where northerly monsoon and trade winds dominate. Continuing past Borneo the path encounters the Intertropical Convergence Zone (ITCZ), the Earth's "weather equator," where northerly and southerly monsoons converge and thunderstorms hold court. Finally, leaving Indonesia and the last island sites, the eclipse track heads out into the Pacific, moving into the variable southerlies along the equator.

Each of these wind and weather regimes has its own peculiarities. However, it is safe to generalize that the eclipse path begins with good prospects of sunshine and ends with cloudier skies. This eclipse is a short one, offering barely two minutes of totality at best, and only one minute in areas with the sunniest weather prospects. Eclipse observers will have to make a difficult choice between eclipse duration and weather when they pick their viewing sites.

THE MIDDLE EAST

Weather systems moving across the shadow's path through Iran and Afghanistan must first cross a protective barrier of mountains that guard the western and northern approaches. These peaks, reaching over 4000 metres high, wring much of the moisture from the air and bring a sunny, semi-arid climate to the interior of these countries. Figure 8 shows that the mean cloud cover for the month ranges between 20 and 40 percent, with lower values found toward the east in Afghanistan. Table 16 mimics this pattern, showing that nearly two mornings out of three are sunny over Iran, but nearly all October mornings are sunny in Afghanistan. This thirsty region is one of the sunniest in the world at this time of year.

Eclipse morning will require a very cloud free sky in the direction of the Sun since the low altitude of the solar disk will aggravate the effects of even a small amount of cloud. The area is very gritty, since October is a windy month, with persistent northwesterlies raising occasional clouds of dust and sand, especially when reinforced by passing cold fronts. Luckily, dust tends to settle overnight when winds decrease and the air stabilizes so the early morning eclipse should occur at the cleanest time of day.

THE INDIAN SUBCONTINENT

During the hot summer months, India is gripped by a humid southeast flow that brings extensive cloudiness and prodigious amounts of precipitation. This southerly monsoon is caused by the heating of the Asian land mass by the summer Sun, drawing air inland from the oceans. The southeast monsoon is essential to the agriculture of the area, though it comes with mixed blessings because of flooding and oppressive humidity.

By October the Sun is well into its southerly winter decline and the Asian land mass is cooling rapidly during the longer nights. The southerly monsoon weakens, and a large anticyclone develops over Tibet and Siberia. The southerlies retreat, to be replaced by a dryer and cooler northerly outflow which arrives after descending the slopes of the Himalayan massif.

The retreat of the southeast monsoon and its replacement by the northerly monsoon begins in September in northern Pakistan and progresses steadily eastward across India through October. By eclipse time, the southern monsoon has usually been pushed into the Bay of Bengal, past Calcutta, and the entire eclipse path is immersed in the drier air flow. The dryness in the air and its origin over the Himalayas also brings cleaner skies for the eclipse.

Across Pakistan and northwestern India, where the northern monsoon has been underway for over a month, mean cloud cover (from satellite observations) falls below 10 percent (Figure 8). Only the interior of the Sahara Desert has less cloudiness in October, and there are few eclipse sites that will ever have better weather prospects than the Great Indian Desert along the Pakistan border.

Surface climatological observations from stations through Pakistan and India show the same pattern of cloudiness (Table 16) as that revealed by the satellites. At Quetta and Jacobabad in Pakistan, mornings are sunny on 25 to 28 days of the month (other stations in Table 16 have too short a period of record to be reliable). This sunny record continues across northwest India, past Delhi and Agra, before declining beyond Allahabad toward Calcutta. Figure 10 shows that mean daily sunshine ranges between 9 and 10 hours over northwest India (out of about 11 possible), and then declines sharply to 7 hours at Calcutta. This pattern is also evident in the October statistics in Table 16.

Since these measurements in the figures and tables are gathered for the whole month of October, and the eclipse occurs close to the end of the month, the numbers favor those areas in the northwest which have been sunny for most or all of the 31 days. The cloudy southwest monsoon tends to withdraw from the Calcutta area during the latter half of October, so statistics for the city represent conditions which are partly characterized by the south monsoon and partly by the north. Thus, Calcutta will appear to be too cloudy in the statistics because of its late change to the northwest monsoon. The November figures might provide a better idea of the cloud prospects.

November sunshine statistics, which represent conditions entirely within the northwest monsoon season, show that sunshine in the Calcutta area increases to about 8.5 hours per day, while northwest India declines slightly to 9.5 hours. Thus the difference between the Calcutta area and the Delhi area at the end of October is not as large as the tables and figures indicate. Sunshine hours are probably about 15 percent higher in the northwest of India than along the Bay of Bengal. Of course all of this depends on the southwest monsoon having left the Calcutta area by eclipse time, a likely but not certain event.

Those in quest of the longest possible eclipse, without sacrificing too much of the good weather, might prefer the Calcutta area to Delhi and Agra. In the latter cities the eclipse duration is less than a minute; a site near Calcutta will see over 20 seconds more of totality. But monsoons are notoriously fickle creatures, and the eclipse date is barely a week after the normal withdrawal of the southerly monsoon at Calcutta. Observers who wish to take a position near the Bay of Bengal should rely on local weather reports and the 1995 status of the northwest monsoon to adjust their viewing plans.

At the infrequent times (15 to 25%) when the path through India is cloudy, the clouds come mostly from high level disturbances caught in the upper atmospheric winds. During summer months, the jet stream resides on the north side of the Himalayas carrying disturbances across Tibet and China. In October, as the continent cools, the jet makes a sudden shift southward, moving to the equatorial side of the mountains. Disturbances which were formerly carried into Tibet now move across northern Pakistan, India and Southeast Asia. Each of these can bring greater or lesser amounts of cloud according to their intensity.

These westerly disturbances are characterized by low pressure systems which move over northern Pakistan and head east northeastward into India. There are about seven or eight of these a month at their peak in January, but there are far fewer in October. The abundance of sunshine provides ample evidence that these systems are not particularly common in the fall.

Morning fog patches are rare across northern India, but increase in frequency toward the coast at Calcutta. Thunderstorms, mostly an afternoon phenomenon, are also more common toward Calcutta. Dust storms are an occasional hazard near the Pakistan border, but rare or unknown elsewhere. Since the northwest monsoon is in its early stages, winds are generally light everywhere and dust is most likely to be raised by the occasional thunderstorm or front.

The Bay of Bengal is famous for its cyclonic storms which occasionally strike the low lying coast of Bangladesh with great loss of life. These storms are not as common as their newspaper reputation would suggest, occurring with a frequency of less than once per month in October. About 30% grow to become severe storms with winds in excess of 90 km/h. Their normal path is to follow the east coast of India, curving to move inland across Bangladesh or northern Myanmar. However, a substantial fraction are not content to follow the common path, and instead head northwestward past Calcutta and into the center of the Indian sub-continent. While they weaken rapidly once they leave their ocean source, the extensive cloudiness thrown up by these storms has the potential to affect eclipse observing as far west as Allahabad, or even a little beyond. Similar cyclonic storms develop in the Arabian Sea, but are very unlikely to reach the eclipse track south of Delhi.

SOUTHEAST ASIA

Across continental Southeast Asia, October is a transitional month with the retreating southerly monsoon being replaced by a light northerly flow. Unfortunately, these northerlies are a part of the trade wind flow from the Pacific and not a piece of the drier air masses which are building over Asia (that will come later in the winter). They carry much more moisture as a result of their trajectory over the warm ocean. These moist winds bring weather patterns which are varied and changeable, and in much of Cambodia, Thailand and Vietnam, October is the wettest month of the year.

The trade wind flow is very light and readily deflected by the many mountain chains and valleys which characterize continental Southeast Asia. Precipitation and cloudiness is highest on the exposed windward side of the terrain, particularly on the east slopes of the coastal mountains of Vietnam, that first intercept the ocean wind flow. The mountain chain that extends the length of Myanmar and forms the backbone of the Malay Peninsula is affected in the same way. In the protected interior of Cambodia and Thailand, however, sunshine is a little more abundant.

Air masses across Southeast Asia are very unsettled and showers and thundershowers grow at the least provocation. It may take little more than two wind flows that collide, a dark surface that collects a little more heat from the Sun, or a range of hills, to set off the buildups. Nearly every afternoon is dotted with clouds, and nights are humid. As in India, the westerly jet also carries upper disturbances into Southeast Asia, each bringing its own retinue of cloud and rain. These systems tend to pass to the north of the eclipse track, except over Myanmar, but are variable enough that they contribute substantially to the cloud cover along the shadow's path.

Heading southeastward past Vietnam, the track encounters the Intertropical Convergence Zone (ITCZ), where the winds from the northerly trades converge with those of the retreating southerly monsoon. This boundary is usually found lurking over the Malay Peninsula, past the northern tip of Borneo and into the Pacific along the 8th parallel. It is not a well-defined boundary, but rather a diffuse area of showers and thunderstorms that wonders about its average position. The land masses of Indochina distort the wind flows that create the ITCZ, and it becomes very broad and undefined between the Bay of Bengal and New Guinea.

Many of the thunderstorms that develop along the ITCZ are part of organized clusters (Figure 9) that throw up large and solid cloud umbrellas. They will almost certainly make eclipse viewing impossible should they arrive on the critical day, and because of their size, will be difficult to outrun. Over the Bay of Bengal, about 2 to 3 of these weather systems can be expected during the month. This grows to 4 to 6 per month off the coast of Vietnam and to 8 along the west coast of Borneo.

Past Borneo and the islands of the Celebes Sea, the path moves into the light southerly flows found on the south side of the ITCZ. Cloud cover decreases slightly north and east of the island of Sulawesi (Celebes), in most part because thunderstorm clusters avoid the area. From a climatological point of view, this is probably the best location for observing from a ship, but a timely weather forecast for other areas would probably give more advantage.

October is in the midst of the typhoon season, a particularly vicious type of storm that is essentially a Pacific hurricane. They have every bit of the nasty personality, and then some, of their North American cousins. Typhoons approach the coasts of Indochina from the east, passing over the Philippines on their way toward Vietnam. They weaken rapidly once they cross onto land, but leave large areas of cloud to plague eclipse-viewing. Luckily, they are quite rare at the latitude of the eclipse track, affecting it perhaps one year in five.

Beyond the Celebes Sea, the eclipse track moves into quieter and slightly sunnier weather of the Pacific equator, where the path ends at sunset in Micronesia. It is a region of cumulus cloudiness, with plenty of blue sky between the convective buildups. Occasional thunderstorm clusters may spoil this idyllic pattern, but they are not as common as farther north toward the ITCZ. Unfortunately, because the eclipsed Sun will be low in the sky, even a little cloudiness will go a long way to obscuring the view.

Shipboard observers will find compatible wave heights along much of the eclipse track, unless a recent typhoon or cyclone sends a large swell into the eclipse area. From the coast of Vietnam to New Guinea wave heights average between one-half and one metre. Beyond New Guinea, and on the Bay of Bengal, waves average a little over one metre in height.

WEATHER SUMMARY

With all of these weather demons converging on Southeast Asia and the islands of Indonesia, the prospects for eclipse observing would seem to be quite dismal. Figure 8 shows that the mean cloudiness ranges from 60 to 80 percent through the area, four to five times higher than over western India. Figure 9 hints at the disadvantage of the close proximity of the ITCZ and its thunderstorm clusters. Table 16, derived from actual observations along the path, also provides a pessimistic outlook with a very low frequency of sunny skies. So what are the actual prospects for clear skies at eclipse time? The best statistic, unfortunately collected at only a few locations, is the actual number of hours of sunshine. Figure 10 promises between 7 and 10 hours at Indian sites. Phnom Penh and Kratie, both in Cambodia, average 6 hours per day in October, and Ho Chi Minh City (Saigon) reports 4.5 (Table 16).

By dividing the number of sunny hours by the number of hours between sunrise and sunset we can calculate the percent of possible sunshine for each site. Such a statistic allows the various locations along the track to be compared directly, and it provides a good estimate of the actual probability of seeing the eclipse except for morning/afternoon differences. It also corrects for the different lengths of the day at the various locations.

An eclipse site in Cambodia or Thailand would have about a 50% probability of seeing the eclipse, in Vietnam about 40%, and locations in western India 85 to 90%. Other values, where available, are given in Table 16. Mobility would normally raise the likelihood by a few percentage points for most eclipses, but the very narrow track for this eclipse limits the distance that can be explored for holes in the cloud. The advantage gained is probably not more than 5%.

One exception is for those who observe at sea where a ship is free to range along the track. Ships are limited by a relatively slow speed, but are not constrained by orographic barriers (islands, perhaps) or the need for roads. Knowing where to sail will be critical, so the advantage to be gained will be determined by forecast accuracy and lead time. There are many variables to consider, but a rough estimate might be that a shipboard chase would confer a 10% advantage.

Unfortunately, sunshine statistics are not available for Indonesian and Philippine islands along the track, and in any event land-based measurements do not accurately reflect conditions on the ocean. We are left with the need to accept substitutes, for which the cloud cover data of Figure 8 will have to suffice. A comparison between locations with similar cloud statistics forces the conclusion that sunshine hours on the South China Sea between Borneo and Vietnam are comparable to conditions at Ho Chi Minh City, and the Celebes Sea near Sulawesi is comparable to the sheltered interior of Cambodia and Thailand. Add 10% for mobility, and shipboard observers have a 50 to 60% probability of suitable skies for the shadow passage.

OBSERVING THE ECLIPSE

EYE SAFETY DURING SOLAR ECLIPSES

The Sun can be viewed safely with the naked eye only during the few brief seconds or minutes of a *total* solar eclipse. Partial and annular solar eclipses are *never* safe to watch without taking special precautions. Even when 99% of the Sun's surface is obscured during the partial phases, the remaining photospheric crescent is intensely bright and cannot be viewed safely without eye protection [Chou, 1981; Marsh, 1982]. *Do not attempt to observe the partial or annular phases of any eclipse with the naked eye. Failure to use appropriate filtration may result in permanent eye damage or blindness!*

Generally, the same equipment, techniques and precautions used to observe the Sun outside of eclipse are required [Pasachoff & Covington, 1993; Pasachoff & Menzel, 1992; Sherrod, 1981]. There are several safe methods that may be used to watch the partial phases. The safest of these is projection, in which a pinhole or small opening is used to cast the image of the Sun on a screen placed a half-meter or more beyond the opening. Projected images of the Sun may even be seen on the ground in the small openings created by interlacing fingers, or in the dappled sunlight beneath a leafy tree. Binoculars can also be used to project a magnified image of the Sun on a white card, but you must avoid the temptation of using these instruments for direct viewing.

Direct viewing of the Sun should only be done using filters specifically designed for this purpose. Such filters usually have a thin layer of aluminum, chromium or silver deposited on their surfaces that attenuates both the visible and the infrared energy. Experienced amateur and professional astronomers may use one or two layers of completely exposed and fully developed black-and-white film, provided the film contains a silver emulsion. Since developed color films lack silver, they are unsafe for use in solar viewing. A widely available alternative for safe eclipse viewing is a number 14 welder's glass. However, only mylar or glass filters specifically designed for the purpose should used with telescopes or binoculars.

Unsafe filters include color film, some non-silver black and white film, smoked glass, photographic neutral density filters and polarizing filters. Solar filters designed to thread into eyepieces and often sold with inexpensive telescopes are also dangerous. They should not be used for viewing the Sun at any time since they often crack from overheating. Do not experiment with other filters unless you are certain that they are safe. Damage to the eyes comes predominantly from invisible infrared wavelengths. The fact that the Sun appears dark in a filter or that you feel no discomfort does not guarantee that your eyes are safe. Avoid all unnecessary risks. Your local planetarium or amateur astronomy club is a good source for additional information.

SKY AT TOTALITY

The total phase of an eclipse is accompanied by the onset of a rapidly darkening sky whose appearance resembles evening twilight about 30 or 40 minutes after sunset. The effect presents an excellent opportunity to view planets and bright stars in the daytime sky. Aside from the sheer novelty of it, such observations are useful in gauging the apparent sky brightness and transparency during totality. The Sun is in Virgo and a number of planets and bright stars will be above the horizon for observers within the umbral path. Figure 11 depicts the appearance of the sky during totality as seen from the center line at 3:30 UT, which corresponds to the western coast of Myanmar. Venus is the brightest planet and can actually be observed in broad daylight provided that the sky is cloud free and of high transparency (i.e. - no dust or particulates). During the 1995 eclipse, Venus is located 17° east of the Sun, having recently passed through superior conjunction in mid-August. Look for the planet during the partial phases by first covering the crescent Sun with an extended hand. During totality, it will be virtually impossible to miss Venus since it shines at a magnitude of $m_v=-3.3$. Although two magnitudes fainter, Jupiter will also be well placed 42° east of the Sun and shining at $m_v=-1.4$. Under good conditions, it may be possible to spot Jupiter 5 to 10 minutes before totality. Since this is a morning eclipse for observers along the Asian path, Jupiter will be low in the southeastern sky and will be below the horizon during totality from Iran through India. Only four days past greatest western elongation, Mercury is 18° west of the Sun at $m_v=-0.3$ and should also be an easy target provided skies are clear. The most difficult of the naked eye planets will be Mars ($m_v=+1.3$), appearing 32° east of the Sun between Venus and Jupiter. Saturn is near opposition 138° east of the Sun and will be below the horizon for all observers. Among the brighter stars visible during totality, Spica ($m_v=+0.7$) is located 7° west of the Sun. Other stars to look for include Regulus ($m_v=+1.35$), Arcturus ($m_v=-0.04$) Capella ($m_v=+0.08$) and Procyon ($m_v=+0.38$). East of India, watch for Acrux ($m_v=+1.33$), Gacrux ($m_v=+1.63v$), Alpha and Beta Centauri ($m_v=-0.01$ & $m_v=+0.6v$) all low in the south while Antares ($m_v=+0.9v$) stands low in the east 7° from Jupiter. Sirius ($m_v=-1.46$) may be seen setting in the southwestern sky for observers along the Iran-India segment of the path.

The following ephemeris [using Bretagnon and Simon, 1986] gives the positions of the naked eye planets during the eclipse. **Delta** is the distance of the planet from Earth (A.U.'s), **V** is the apparent visual magnitude of the planet, and **Elong** gives the solar elongation or angle between the Sun and planet. Note that Saturn is near opposition and will be below the horizon for all observers during the eclipse.

```
Planetary Ephemeris:   1995 Oct 24   5:00 UT    Equinox = Mean Date

Planet       RA           Dec          Delta    V     Size    Phase   Elong
           h  m  s      °   '   "                     "                 °
Sun        13 52 50    -11-34-47      0.99476  -26.7  1929.4    -       -
Mercury    12 50 26     -3 -9-41      1.07410   -0.6     6.3   0.69   17.6W
Venus      14 59 26    -16-50-37      1.61647   -3.3    10.3   0.96   17.0E
Mars       16  0 43    -21-18-40      2.21358    1.6     4.2   0.97   32.1E
Jupiter    16 52 51    -22-10-18      5.96082   -1.4    33.0   1.00   44.2E
Saturn     23 22  2     -6-36-26      8.83178    0.1    18.7   1.00  138.3E
```

ECLIPSE PHOTOGRAPHY

The eclipse may be safely photographed provided that the above precautions are followed. Almost any kind of camera with manual controls can be used to capture this rare event. However, a lens with a fairly long focal length is recommended to produce as large an image of the Sun as possible. A standard 50 mm lens yields a minuscule 0.5 mm image, while a 200 mm telephoto or zoom produces a 1.9 mm image. A better choice would be one of the small, compact catadioptic or mirror lenses that have become widely available in the past ten years. The focal length of 500 mm is most common among such mirror lenses and yields a solar image of 4.6 mm. Adding 2x tele-converter will produce a 1000 mm focal length, which doubles the Sun's size to 9.2 mm. Focal lengths in excess of 1000 mm usually fall within the realm of amateur telescopes. If full disk photography of partial phases on 35 mm format is planned, the focal length of the telescope or lens must be 2600 mm or less. Longer focal lengths will only permit photography of a portion of the Sun's disk. Furthermore, in order to photograph the Sun's corona during totality, the focal length should be no longer than 1500 mm to 1800 mm (for 35 mm equipment). For any particular focal length, the diameter of the Sun's image is approximately equal to the focal length divided by 109.

A mylar or glass solar filter must be used on the lens throughout the partial phases for both photography and safe viewing. Such filters are most easily obtained through manufacturers and dealers listed in *Sky & Telescope* and *Astronomy* magazines. These filters typically attenuate the Sun's visible and infrared energy by a factor of 100,000. However, the actual filter factor and choice of ISO film speed will play critical roles in determining the correct photographic exposure. A low to medium speed film is recommended (ISO 50 to 100) since the Sun gives off abundant light. The easiest method for determining the correct exposure is accomplished by running a calibration test on the uneclipsed Sun. Shoot a roll of film of the mid-day Sun at a fixed aperture (f/8 to f/16) using every shutter speed between 1/1000 and 1/4 second. After the film is developed, note the best exposures and use them to photograph all the partial phases since the Sun's surface brightness remains constant throughout the eclipse. The exposure should also be increased by one or two stops for narrow crescent phases to compensate for limb darkening. Bracketing of two or more stops may be necessary if hazy skies or clouds interfere on eclipse day.

Certainly the most spectacular and awe inspiring phase of the eclipse is totality. For a few brief minutes or seconds, the Sun's pearly white corona, red prominences and chromosphere are visible. The great challenge is to obtain a set of photographs which capture some aspect of these fleeting phenomena. The most important point to remember is that during the total phase, all solar filters *must be removed!* The corona has a surface brightness a million times fainter than the photosphere, so photographs of the corona are made without a filter. Furthermore, it is completely safe to view the totally eclipsed Sun directly with the naked eye. No filters are needed and they will only hinder your view. The average brightness of the corona varies inversely with the distance from the Sun's limb. The inner corona is far brighter than the outer corona. Thus, no one exposure can capture its the full dynamic range. The best strategy is to choose one aperture or f/number and bracket the exposures over a range of shutter speeds (i.e. - 1/1000 down to 1 second). Rehearsing this sequence is highly recommended since great excitement accompanies totality and there is little time to think.

Exposure times for various combinations of film speeds (ISO), apertures (f/number) and solar features (chromosphere, prominences, inner, middle and outer corona) are summarized in Table 17. The table was developed from eclipse photographs made by Espenak as well as from photographs published in *Sky and Telescope*. To use the table, first select the ISO film speed in the upper left column. Next, move to the right to the desired aperture or f/number for the chosen ISO. The shutter speeds in that column may be used as starting points for photographing various features and phenomena tabulated in the 'Subject' column at the far left. For example, to photograph prominences using ISO 100 at f/11, the table recommends an exposure of 1/500. Alternatively, you can calculate the recommended shutter speed using the 'Q' factors tabulated along with the exposure formula at the bottom of Table 17. Keep in mind that these exposures are based on a clear sky and an average corona. You should bracket your exposures one or more stops to take into account the actual sky conditions and the variable nature of these phenomena.

Another interesting way to photograph the eclipse is to record its various phases all on one frame. This is accomplished by using a stationary camera capable of making multiple exposures (check the camera instruction manual). Since the Sun moves through the sky at the rate of 15 degrees per hour, it slowly drifts through the field of view of any camera equipped with a normal focal length lens (i.e. - 35 to 50 mm). If the camera is oriented so that the Sun drifts along the frame's diagonal, it will take over three hours for the Sun to cross the field of a 50 mm lens. The proper camera orientation can be determined through trial and error several days before the eclipse. This will also insure that no trees or buildings obscure the camera's view during the eclipse. The Sun should be positioned along the eastern (right in southern hemisphere) edge or corner of the viewfinder shortly before the eclipse begins. Exposures are then made throughout the eclipse at five minute intervals. The camera must remain perfectly rigid during this period and may be clamped to a wall or fence post since tripods are easily bumped. The final photograph will consist of a string of Suns, each showing a different phase of the eclipse.

Finally, an eclipse effect that is easily captured with point-and-shoot or automatic cameras should not be overlooked. During the eclipse, the ground under nearby shade trees is covered with small images of the crescent Sun. The gaps between the tree leaves act like pinhole cameras and each one projects its own tiny image of the Sun. The effect can be duplicated by forming a small aperture with one's hands and watching the ground below. The pinhole camera effect becomes more prominent with increasing eclipse magnitude. Virtually any camera can be used to photograph the phenomenon, but automatic cameras must have their flashes turned off since this would otherwise obliterate the pinhole images.

For more information on eclipse photography, observations and eye safety, see FURTHER READING in the BIBLIOGRAPHY.

CONTACT TIMINGS FROM THE PATH LIMITS

Precise timings of beading phenomena made near the northern and southern limits of the umbral path (i.e. - the graze zones), are of value in determining the diameter of the Sun relative to the Moon at the time of the eclipse. Such measurements are essential to an ongoing project to monitor changes in the solar diameter. Due to the conspicuous nature of the eclipse phenomena and their strong dependence on geographical location, scientifically useful observations can be made with relatively modest equipment. A small telescope, short wave radio and portable camcorder are usually used to make such measurements. Time signals are broadcast via short wave stations WWV and CHU, and are recorded simultaneously as the eclipse is videotaped. If a video camera is not available, a tape recorder can be used to record time signals with verbal timings of each event. Inexperienced observers are cautioned to use great care in making such observations. The safest timing technique consists of observing a projection of the Sun rather than directly imaging the solar disk itself. The observer's geodetic coordinates are required and can be measured from USGS or other large scale maps. If a map is unavailable, then a detailed description of the observing site should be included which provides information such as distance and directions of the nearest towns/settlements, nearby landmarks, identifiable buildings and road intersections. The method of contact timing should be described in detail, along with an estimate of the error. The precisional requirements of these observations are ±0.5 seconds in time, 1" (~30 meters) in latitude and longitude, and ±20 meters (~60 feet) in elevation. Although GPS's (Global Positioning Satellite receivers) are commercially available (~$500 US), their positional accuracy of ±100 meters is about three times larger than the minimum accuracy required by grazing eclipse measurements. The International Occultation Timing Association (IOTA) coordinates observers world-wide during each eclipse. For more information, contact:

Dr. David W. Dunham, IOTA
7006 Megan Lane
Greenbelt, MD 20770-3012   Phone: (301) 474-4722
U. S. A.   Internet: David_Dunham@jhuapl.edu

Send reports containing graze observations, eclipse contact and Baily's bead timings, including those made anywhere near or in the path of totality or annularity to:

Dr. Alan D. Fiala
Orbital Mechanics Dept.
U. S. Naval Observatory
3450 Massachusetts Ave., NW
Washington, DC 20392-5420

PLOTTING THE PATH ON MAPS

If high resolution maps of the umbral path are needed, the coordinates listed in Table 7 are conveniently provided at 1° increments of longitude to assist plotting by hand. The path coordinates in Table 3 define a line of maximum eclipse at five minute increments in Universal Time. It is also advisable to include lunar limb corrections to the northern and southern limits listed in Table 6, especially if observations are planned from the graze zones. Global Navigation Charts (1:5,000,000), Operational Navigation Charts (scale 1:1,000,000) and Tactical Pilotage Charts (1:500,000) of many parts of the world are published by the Defense Mapping Agency. In October 1992, the DMA discontinued selling maps directly to the general public. This service has been transferred to the National Ocean Service (NOS). For specific information about map availability, purchase prices, and ordering instructions, contact the NOS at:

National Ocean Service
Distribution Branch
N/GC33
6501 Lafayette Avenue
Riverdale, MD 20737, USA   Phone: 1-301-436-6990

It is also advisable to check the telephone directory for any map specialty stores in your city or metropolitan area. They often have large inventories of many maps available for immediate delivery.

ECLIPSE DATA ON INTERNET

NASA ECLIPSE BULLETINS ON INTERNET

Response to the first two NASA solar eclipse bulletins RP1301 (Annular Solar Eclipse of 10 May 1994) and RP1318 (Total Solar Eclipse of 3 Nov 1994) was overwhelming. Unfortunately, the demand exceeded the supply and current funding sources did not permit additional printings to fill all requests. To address this problem as well as allowing greater access to them, these eclipse bulletins were both made available via the Internet in April 1994. This was due entirely through the kind efforts and expertise of Dr. Joe Gurman (GSFC/Solar Physics Branch). The 1995 bulletin is also being made available via Internet.

Formats include BinHex-encoded versions of the original Microsoft Word (for Macintosh) text files and those figures generated in PICT format, GIF versions of all the figures, and JPEG scans of the GNC maps of the eclipse path, as well as hypertext versions. They can be read or downloaded via the World-Wide Web server with a mosaic client from the SDAC (Solar Data Analysis Center) home page:

> http://umbra.gsfc.nasa.gov/sdac.html

The top-level URL for the eclipse bulletins themselves are:

> http://umbra.gsfc.nasa.gov/eclipse/940510/rp.html   (1994 May 10)
> http://umbra.gsfc.nasa.gov/eclipse/941103/rp.html   (1994 Nov 3)
> http://umbra.gsfc.nasa.gov/eclipse/951024/rp.html   (1995 Oct 24)

BinHex-encoded, StuffIt Lite-compressed version of the original Word and PICT files are available via anonymous ftp at:

> http://umbra.gsfc.nasa.gov/eclipse/940510   (1994 May 10)
> http://umbra.gsfc.nasa.gov/eclipse/941103   (1994 Nov 3)
> http://umbra.gsfc.nasa.gov/eclipse/951024   (1995 Oct 24)

Directories of GIF figures, ASCII tables, and JPEG maps are also accessible through the Web.

Current plans call for making all future NASA eclipse bulletins available over the Internet, at or before publication of each. It should be noted that this is a first attempt to make the bulletins available electronically. The primary goal has been to make the bulletins available to as large an audience as possible. Thus, some figures or maps may not be at their optimum resolution or format. Comments and suggestions are actively solicited to fix problems and improve on compatibility and formats.

FUTURE ECLIPSE PATHS ON INTERNET

Presently, the NASA eclipse bulletins are published 12 to 18 months before each eclipse. This will soon be increased to 24 months or more. However, there have been a growing number of requests for eclipse path data with an even greater lead time. To accommodate the demand, predictions have been generated for all central solar eclipses from 1995 through 2000 using the JPL DE/LE 200 ephemerides. All predictions use the Moon's the center of mass; no corrections have been made to adjust for center of figure. The value used for the Moon's mean radius is k=0.272281. The umbral path characteristics have been predicted at 2 minute intervals of time compared to the 6 minute interval used in *Fifty Year Canon of Solar Eclipses: 1986-2035* [Espenak, 1987]. This should provide enough detail for making preliminary plots of the path on larger scale maps. Note that positive latitudes are north and positive longitudes are west.

The paths for the following seven eclipses are currently available via the Internet:

> 29 Apr 1995 – Annular Solar Eclipse
> 24 Oct 1995 – Total Solar Eclipse
> 9 Mar 1997   – Total Solar Eclipse
> 26 Feb 1998 – Total Solar Eclipse
> 22 Aug 1998 – Annular Solar Eclipse
> 16 Feb 1999 – Annular Solar Eclipse
> 11 Aug 1999 – Total Solar Eclipse

The tables can be accessed with Mosaic through SDAC home page, or directly at URL:

http://umbra.gsfc.nasa.gov/eclipse/predictions/eclipse-paths.html

Comments, corrections or suggestions should be sent to Fred Espenak either via regular or e-mail ("u32fe@lepvax.gsfc.nasa.gov"). For Internet related problems, please contact Joe Gurman ("gurman@uvsp.gsfc.nasa.gov"). Either one of us can also be contacted for more detailed descriptions of formats and directories or instructions for downloading files using Mosaic or ftp.

ALGORITHMS, EPHEMERIDES AND PARAMETERS

Algorithms for the eclipse predictions were developed by Espenak primarily from the *Explanatory Supplement* [1974] with additional algorithms from Meeus, Grosjean and Vanderleen [1966] and Meeus [1982]. The solar and lunar ephemerides were generated from the JPL DE200 and LE200, respectively. All eclipse calculations were made using a value for the Moon's radius of **k**=0.2722810 for umbral contacts, and **k**=0.2725076 (adopted IAU value) for penumbral contacts. Center of mass coordinates were used except where noted. An extrapolated value for ΔT of 61.0 seconds was used to convert the predictions from Terrestrial Dynamical Time to Universal Time.

The primary source for geographic coordinates used in the local circumstances tables is *The New International Atlas* (Rand McNally, 1991). Elevations for major cities were taken from *Climates of the World* (U. S. Dept. of Commerce, 1972).

All eclipse predictions presented in this publication were generated on a Macintosh computer. As such, it represents the culmination of a two year project to migrate a great deal of eclipse software from mainframe (DEC VAX 11/785) to personal computer (Macintosh IIfx) and from one programming language (FORTRAN IV) to another (THINK Pascal). Word processing and page layout for the publication were done using Microsoft Word v5.1. Figure annotation was done with Claris MacDraw Pro 1.5. All meteorological diagrams were prepared using Windows Draw 3.0 and converted to Macintosh format.

The names and spellings of countries, cities and other geopolitical regions are not authoritative, nor do they imply any official recognition in status. Corrections to names, geographic coordinates and elevations are actively solicited in order to update the data base for future eclipses. All calculations, diagrams and opinions presented in this publication are those of the authors and they assume full responsibility for their accuracy.

BIBLIOGRAPHY

REFERENCES

Bretagnon, P., and Simon, J. L., *Planetary Programs and Tables from –4000 to +2800*, Willmann-Bell, Richmond, Virginia, 1986.
Chou, B. R., "Safe Solar Filters," *Sky & Telescope*, August 1981, p. 119.
Climates of the World, U. S. Dept. of Commerce, Washington DC, 1972.
Dunham, J. B, Dunham, D. W. and Warren, W. H., *IOTA Observer's Manual*, (draft copy), 1992.
Espenak, F., *Fifty Year Canon of Solar Eclipses: 1986–2035*, NASA RP-1178, Greenbelt, MD, 1987.
Explanatory Supplement to the Astronomical Ephemeris and the American Ephemeris and Nautical Almanac, Her Majesty's Nautical Almanac Office, London, 1974.
Herald, D., "Correcting Predictions of Solar Eclipse Contact Times for the Effects of Lunar Limb Irregularities," *J. Brit. Ast. Assoc.*, 1983, **93**, 6.
Marsh, J. C. D., "Observing the Sun in Safety," *J. Brit. Ast. Assoc.*, 1982, **92**, 6.
Meeus, J., *Astronomical Formulae for Calculators*, Willmann-Bell, Inc., Richmond, 1982.
Meeus, J., Grosjean, C., and Vanderleen, W., *Canon of Solar Eclipses*, Pergamon Press, New York, 1966.
Morrison, L. V., "Analysis of lunar occultations in the years 1943–1974...," *Astr. J.*, 1979, **75**, 744.
Morrison, L.V., and Appleby, G.M., "Analysis of lunar occultations - III. Systematic corrections to Watts' limb-profiles for the Moon," *Mon. Not. R. Astron. Soc.*, 1981, **196**, 1013.
The New International Atlas, Rand McNally, Chicago/New York/San Francisco, 1991.
van den Bergh, G., *Periodicity and Variation of Solar (and Lunar) Eclipses*, Tjeenk Willink, Haarlem, Netherlands, 1955.
Watts, C. B., "The Marginal Zone of the Moon," *Astron. Papers Amer. Ephem.*, 1963, **17**, 1-951.

FURTHER READING

Allen, D., and Allen, C., *Eclipse*, Allen & Unwin, Sydney, 1987.
Astrophotography Basics, Kodak Customer Service Pamphlet P150, Eastman Kodak, Rochester, 1988.
Brewer, B., *Eclipse*, Earth View, Seattle, 1991.
Covington, M., *Astrophotography for the Amateur*, Cambridge University Press, Cambridge, 1988.
Espenak, F., "Total Eclipse of the Sun," *Petersen's PhotoGraphic*, June 1991, p. 32.
Fiala, A. D., DeYoung, J. A., and Lukac, M. R., *Solar Eclipses, 1991–2000*, USNO Circular No. 170, U. S. Naval Observatory, Washington, DC, 1986.
Harris, J., and Talcott, R.,*Chasing the Shadow*, Kalmbach Pub., Waukesha, 1994.
Littmann, M., and Willcox, K., *Totality, Eclipses of the Sun*, University of Hawaii Press, Honolulu, 1991.
Lowenthal, J., *The Hidden Sun: Solar Eclipses and Astrophotography*, Avon, New York, 1984.
Mucke, H., and Meeus, J., *Canon of Solar Eclipses: –2003 to +2526*, Astronomisches Büro, Vienna, 1983.
North, G., *Advanced Amateur Astronomy*, Edinburgh University Press, 1991.
Oppolzer, T. R. von, *Canon of Eclipses*, Dover Publications, New York, 1962.
Ottewell, G., *The Under-Standing of Eclipses*, Astronomical Workshop, Greenville, NC, 1991.
Pasachoff, J. M., and Covington, M., *Cambridge Guide to Eclipse Photography*, Cambridge University Press, Cambridge and New York, 1993.
Pasachoff, J. M., and Menzel, D. H., *Field Guide to the Stars and Planets*, 3rd edition, Houghton Mifflin, Boston, 1992.
Sherrod, P. C., *A Complete Manual of Amateur Astronomy*, Prentice-Hall, 1981.
Sweetsir, R., and Reynolds, M., *Observe: Eclipses*, Astronomical League, Washington, DC, 1979.
Zirker, J. B., *Total Eclipses of the Sun*, Van Nostrand Reinhold, New York, 1984.

TOTAL SOLAR ECLIPSE OF 1995 OCTOBER 24

FIGURES

Figure 1: **ORTHOGRAPHIC PROJECTION MAP OF THE ECLIPSE PATH**
Total Solar Eclipse of 1995 Oct 24

Geocentric Conjunction = 04:22:31.1 UT   J.D. = 2450014.682304
Greatest Eclipse = 04:32:29.5 UT   J.D. = 2450014.689230
Eclipse Magnitude = 1.02134   Gamma = 0.35176
Saros Series = 143   Member = 22 of 72

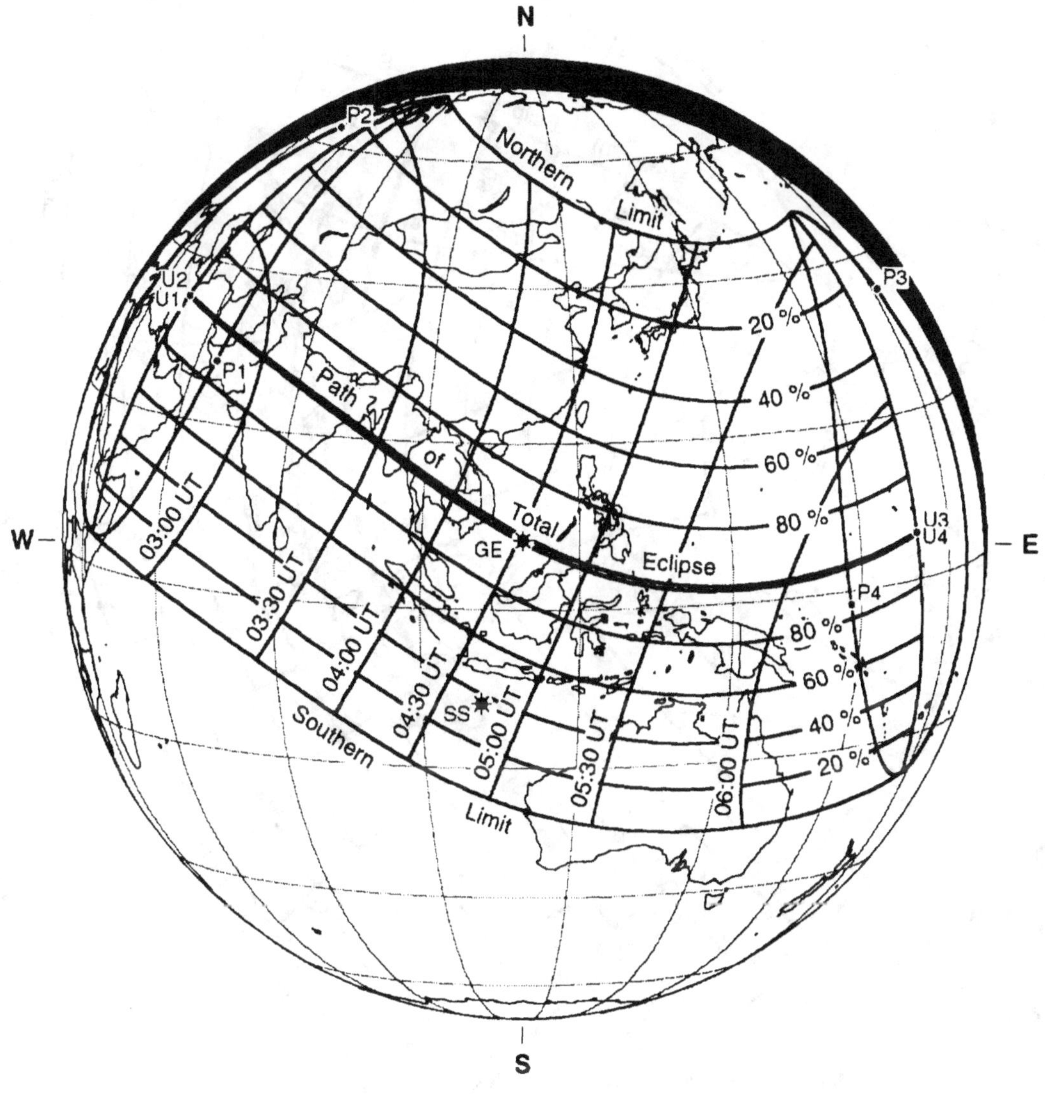

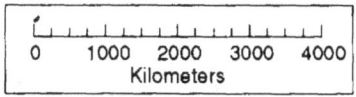

F. Espenak, NASA/GSFC - Mon, Jan 24, 1994

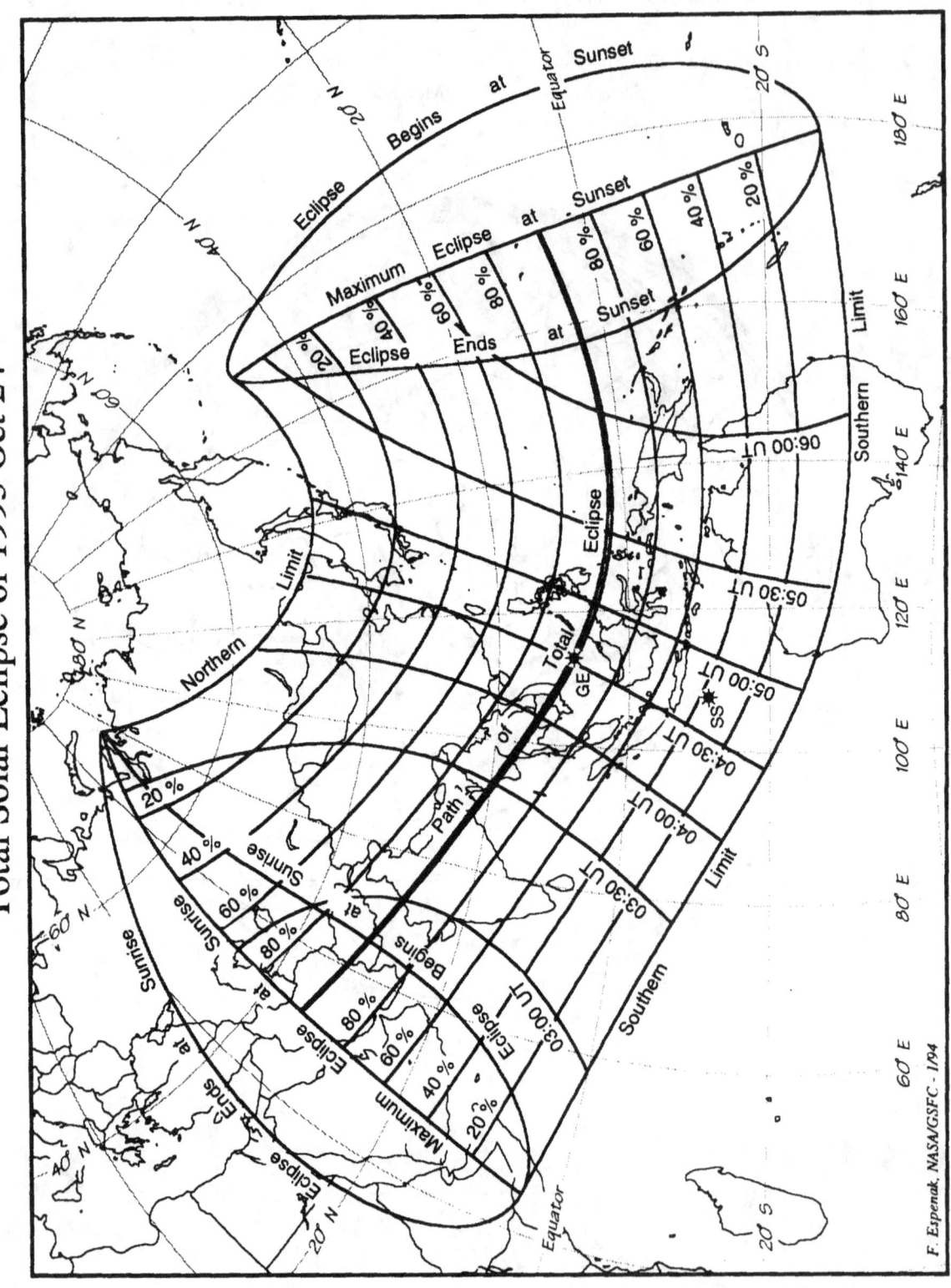

Figure 2: Stereographic Projection Map of The Eclipse Path

Total Solar Eclipse of 1995 Oct 24

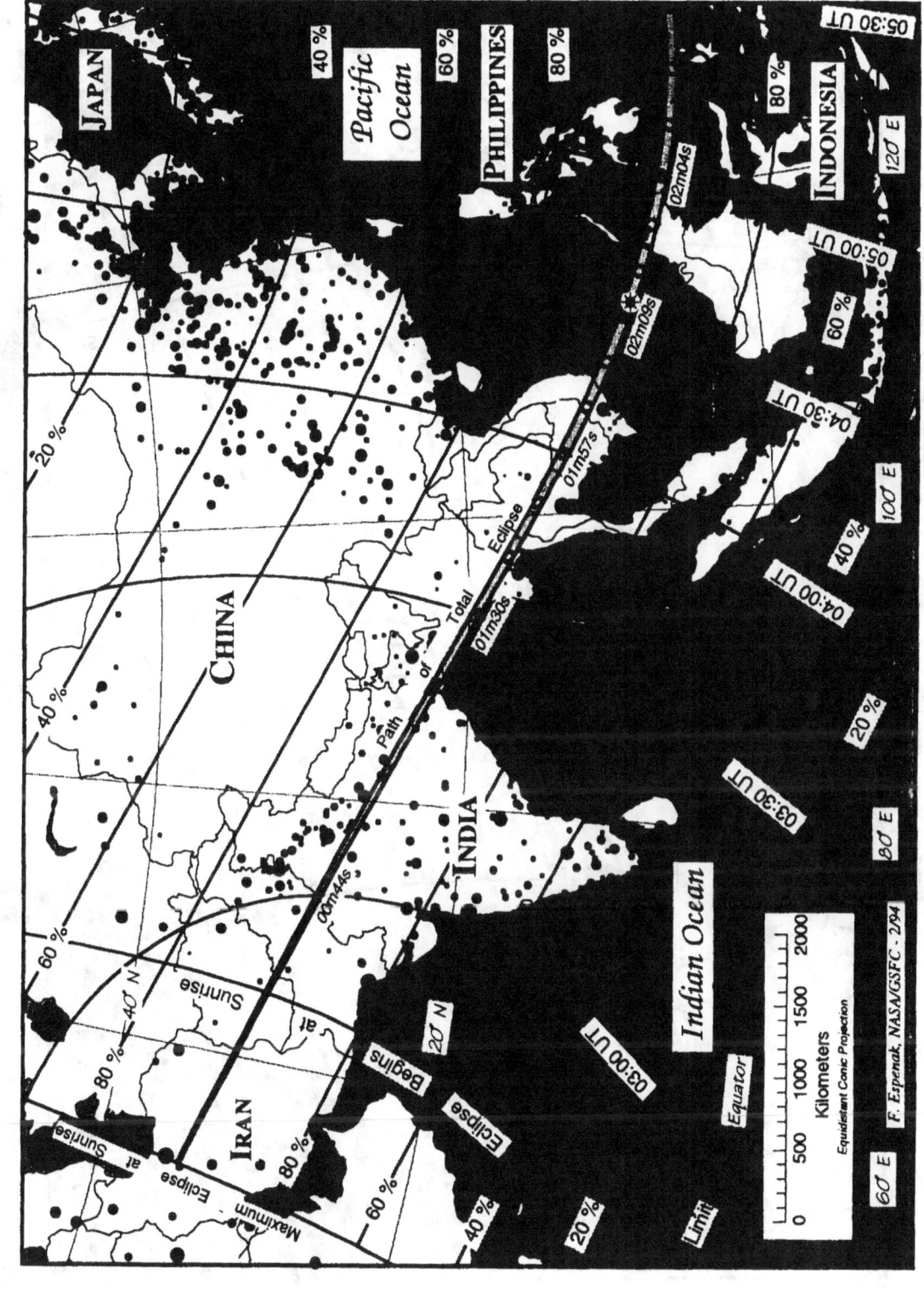

Figure 3: THE ECLIPSE PATH IN ASIA
Total Solar Eclipse of 1995 Oct 24

Figure 4: THE ECLIPSE PATH IN INDIA

Total Solar Eclipse of 1995 Oct 24

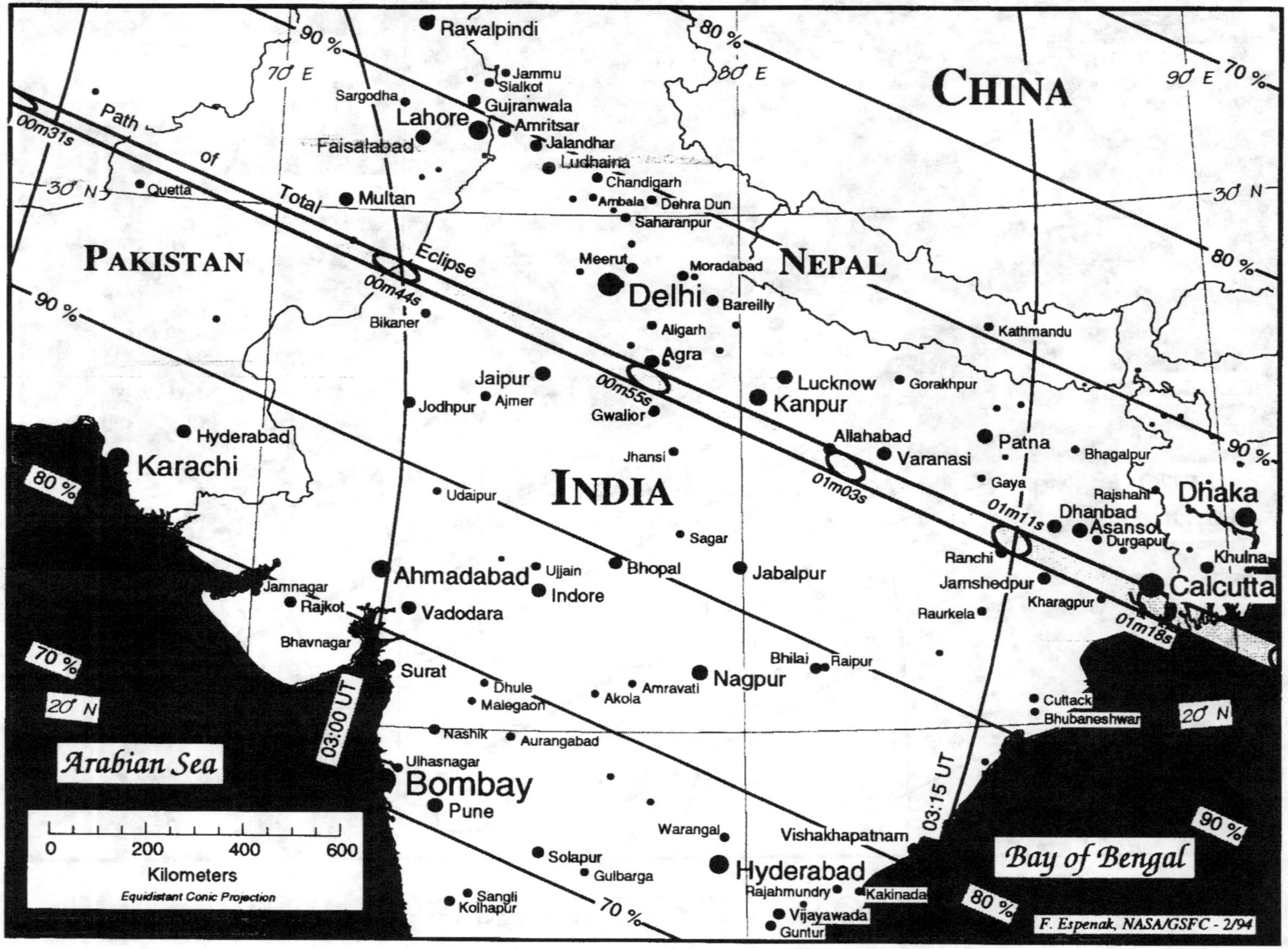

Figure 5: THE ECLIPSE PATH IN SOUTHEAST ASIA

Total Solar Eclipse of 1995 Oct 24

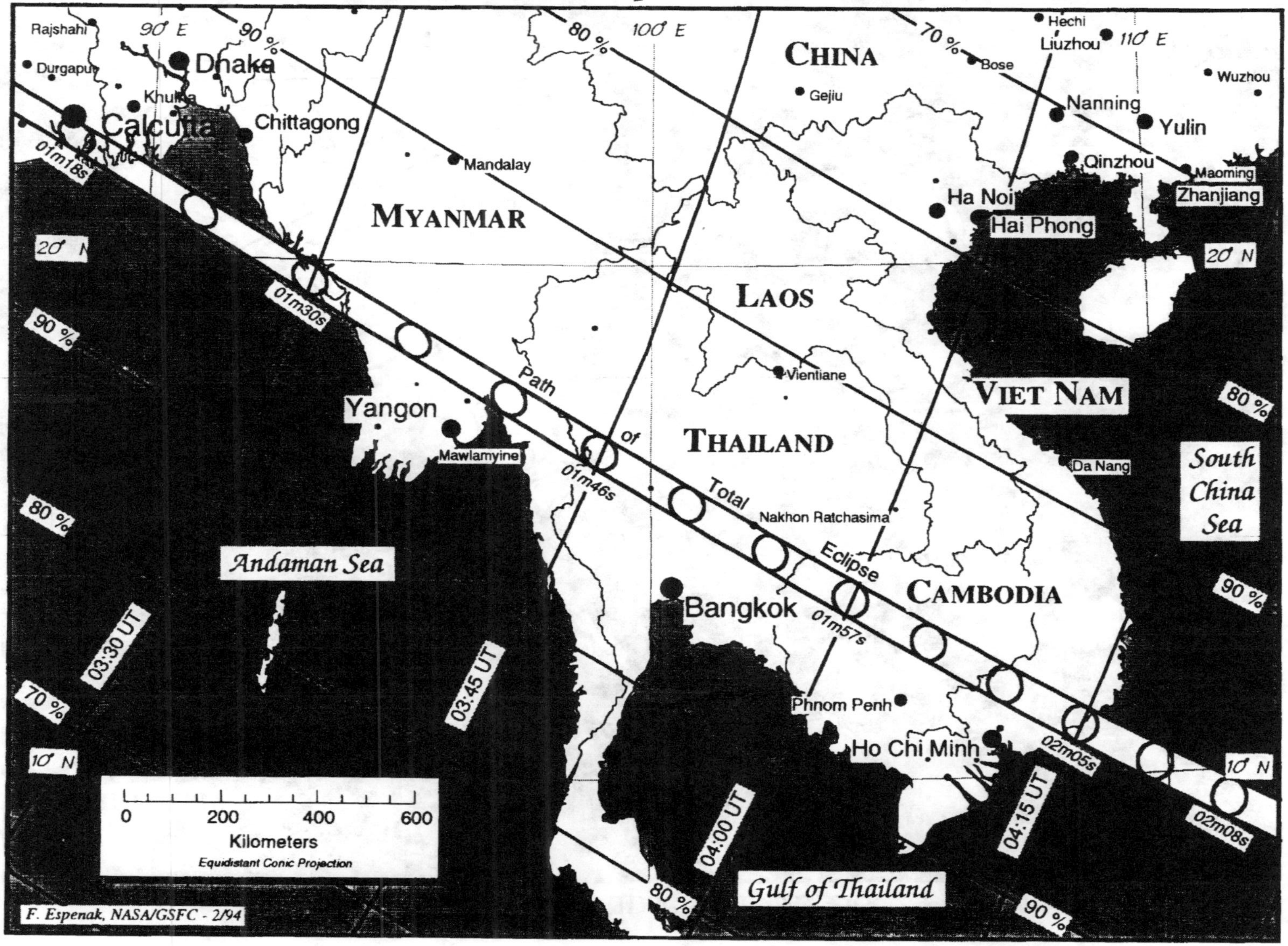

Figure 6: The Eclipse Path in South China Sea and Celebes Sea

Total Solar Eclipse of 1995 Oct 24

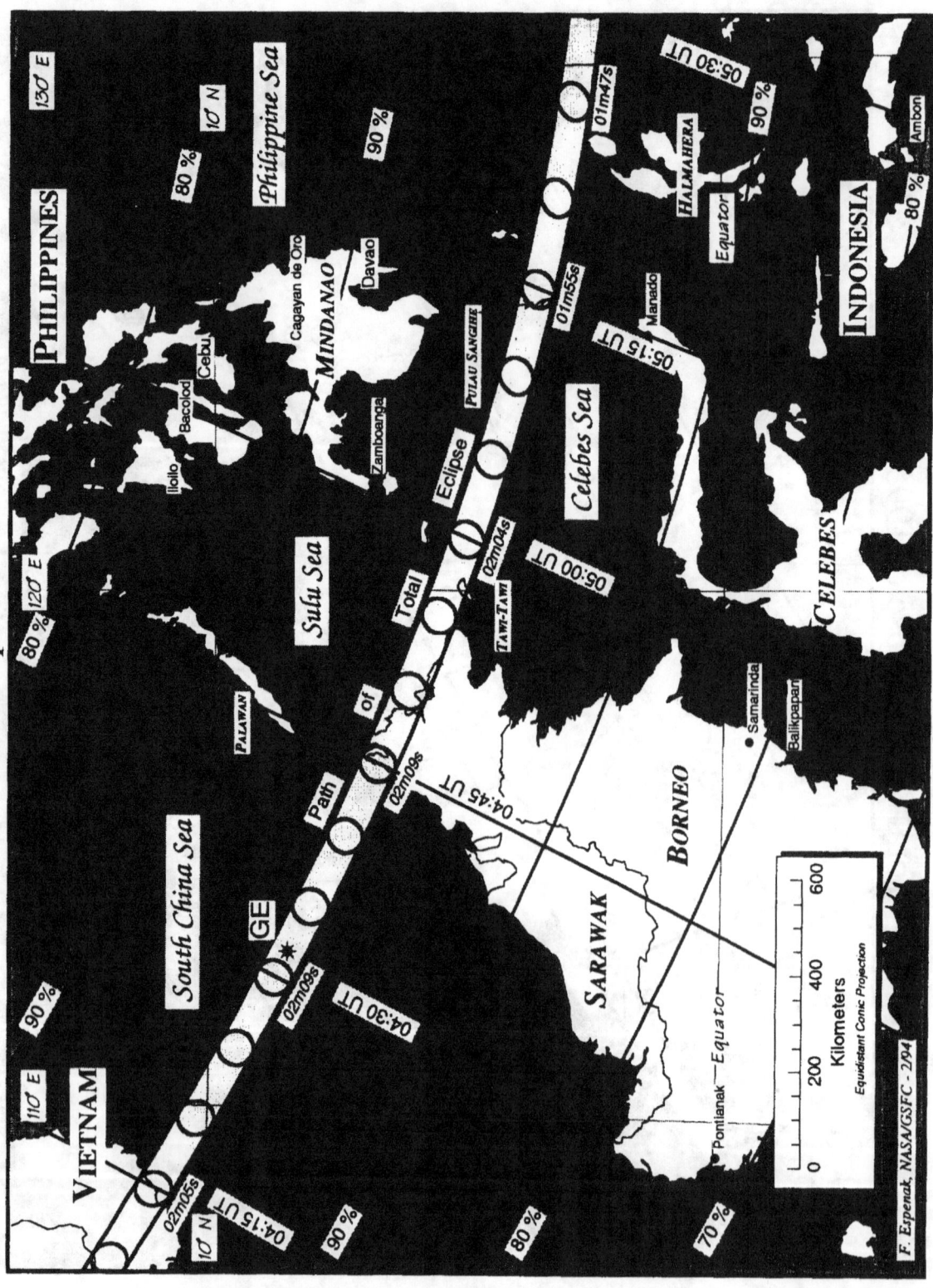

28

Figure 7: **THE LUNAR LIMB PROFILE AT 03:30 UT**

Total Solar Eclipse of 1995 Oct 24

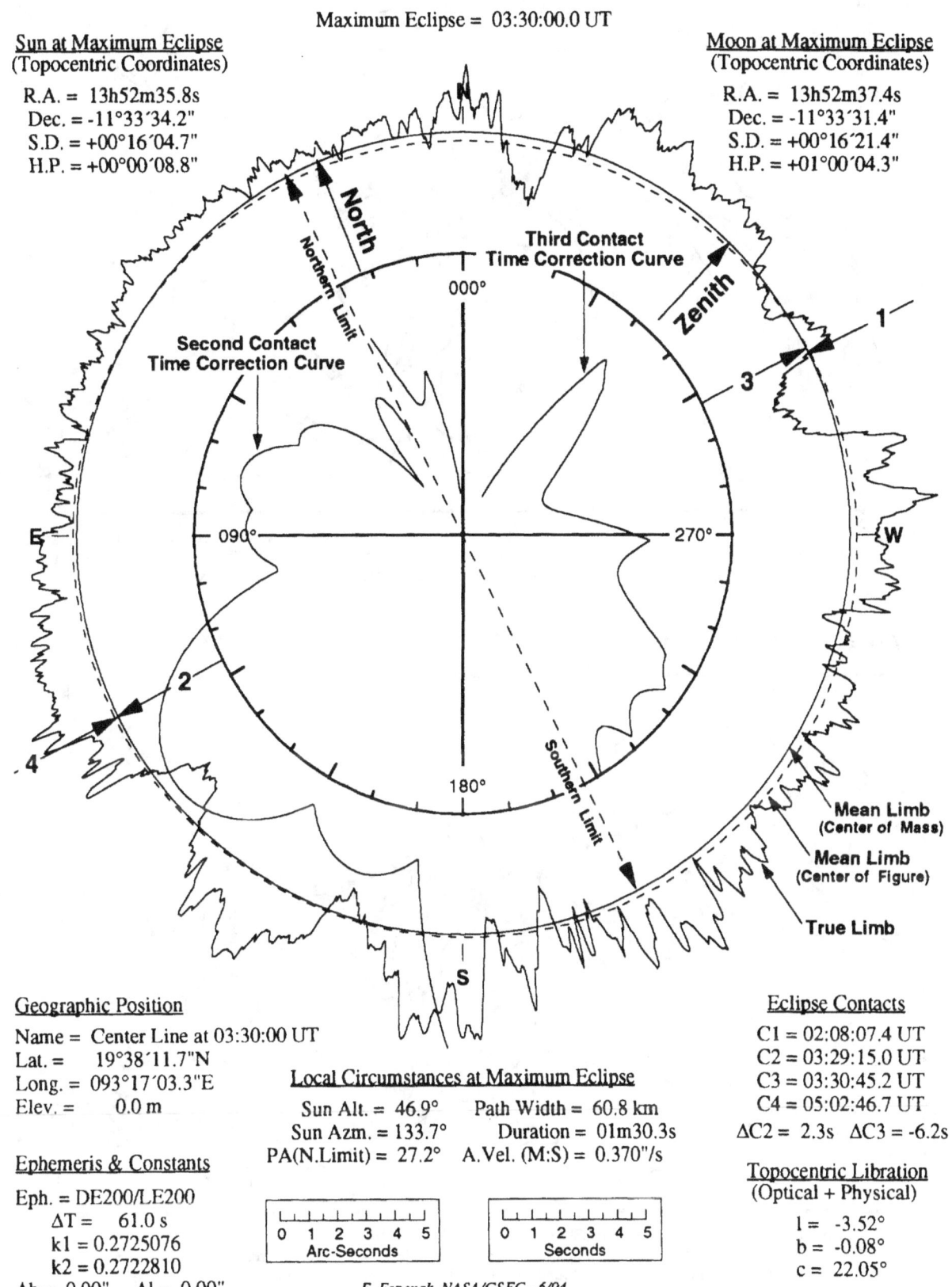

F. Espenak, NASA/GSFC - 6/94

Figure 8: **MEAN CLOUD COVER IN OCTOBER ALONG THE ECLIPSE PATH**

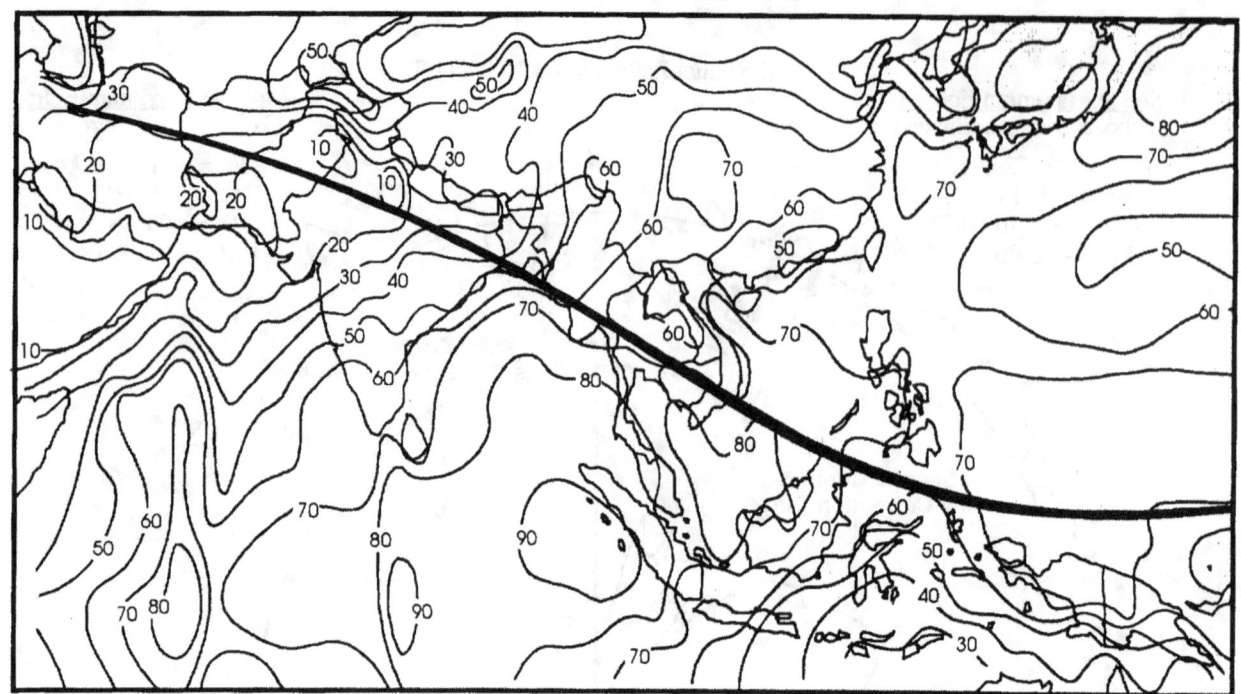

Figure 8: Mean cloud amount (in percent) for October along the eclipse path. This data was extracted from 8 years of satellite data as part of the International Satellite Cloud Climatology Project. Cloud data was collected in 2°x2° degree latitude/longitude bins from a series of satellite images throughout the day. It thus represents a mean over the entire day averaged for the month.

Figure 9: **FREQUENCY OF HIGHLY REFLECTIVE CLOUDS IN OCTOBER ALONG THE ECLIPSE PATH**

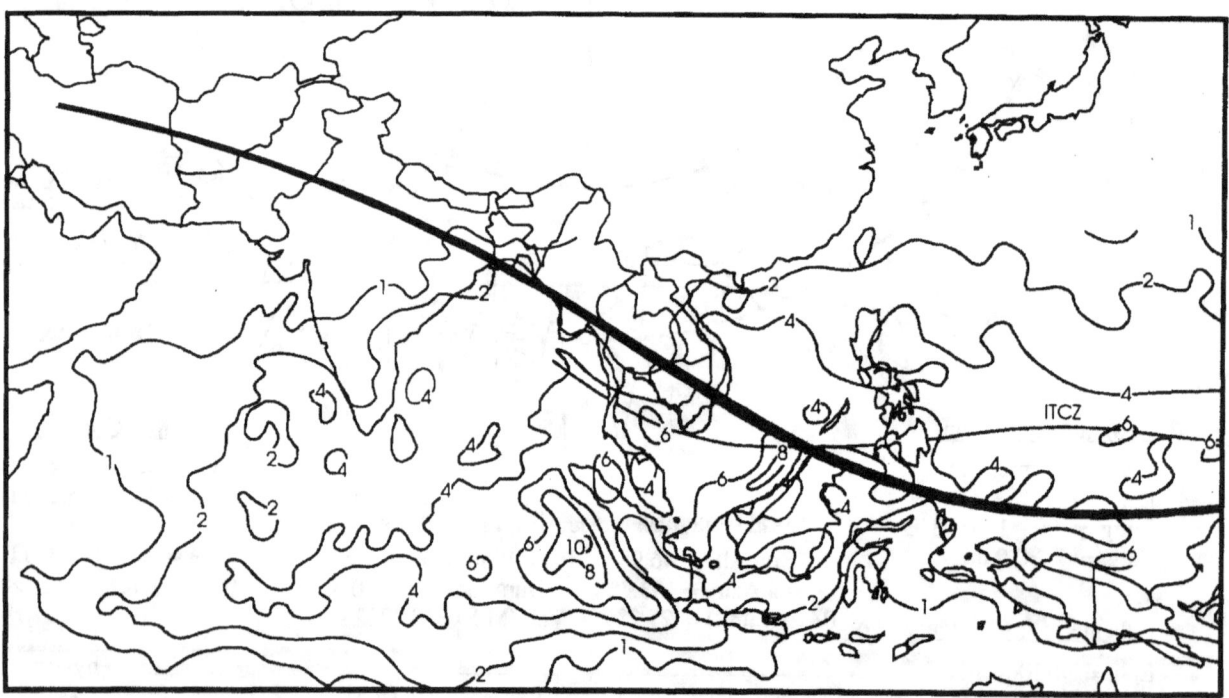

Figure 9: Monthly frequency in October of highly reflective clouds. The HRC dataset uses infrared and visible satellite imagery to measure the frequency of occurrence of large-scale thunderstorm complexes at a 1° spatial resolution. Contours in this figure show the number of days in the month in which large thunderstorm clusters were detected at a particular location. Single or small groups of thunderstorms are not included in the analysis - systems must be at least 200 km across to be incorporated into the dataset. (Garcia, O., 1985, Atlas of Highly Reflective Clouds of the Global Tropics, U.S. Department of Commerce, NTIS PB-87129169)

Figure 10: MEAN NUMBER OF HOURS OF SUNSHINE IN OCTOBER FOR INDIA

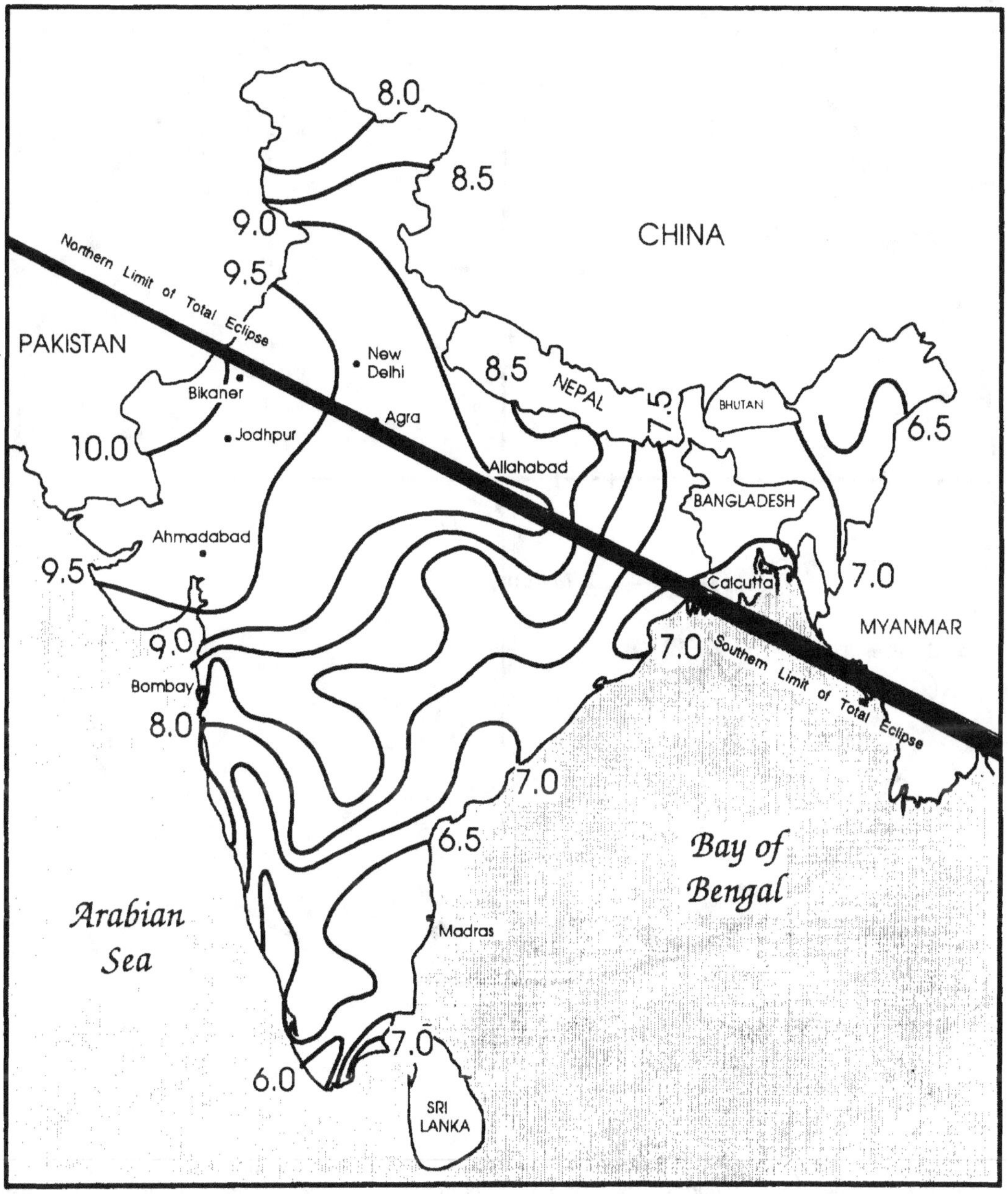

Figure 10: Mean number of hours of sunshine for the month of October in India. Small differences between this chart and the data in table 1 is likely due to different periods of record and the smoothing inherent in constructing a contour map.

Figure 11: THE SKY DURING TOTALITY AS SEEN FROM CENTER LINE AT 03:30 UT

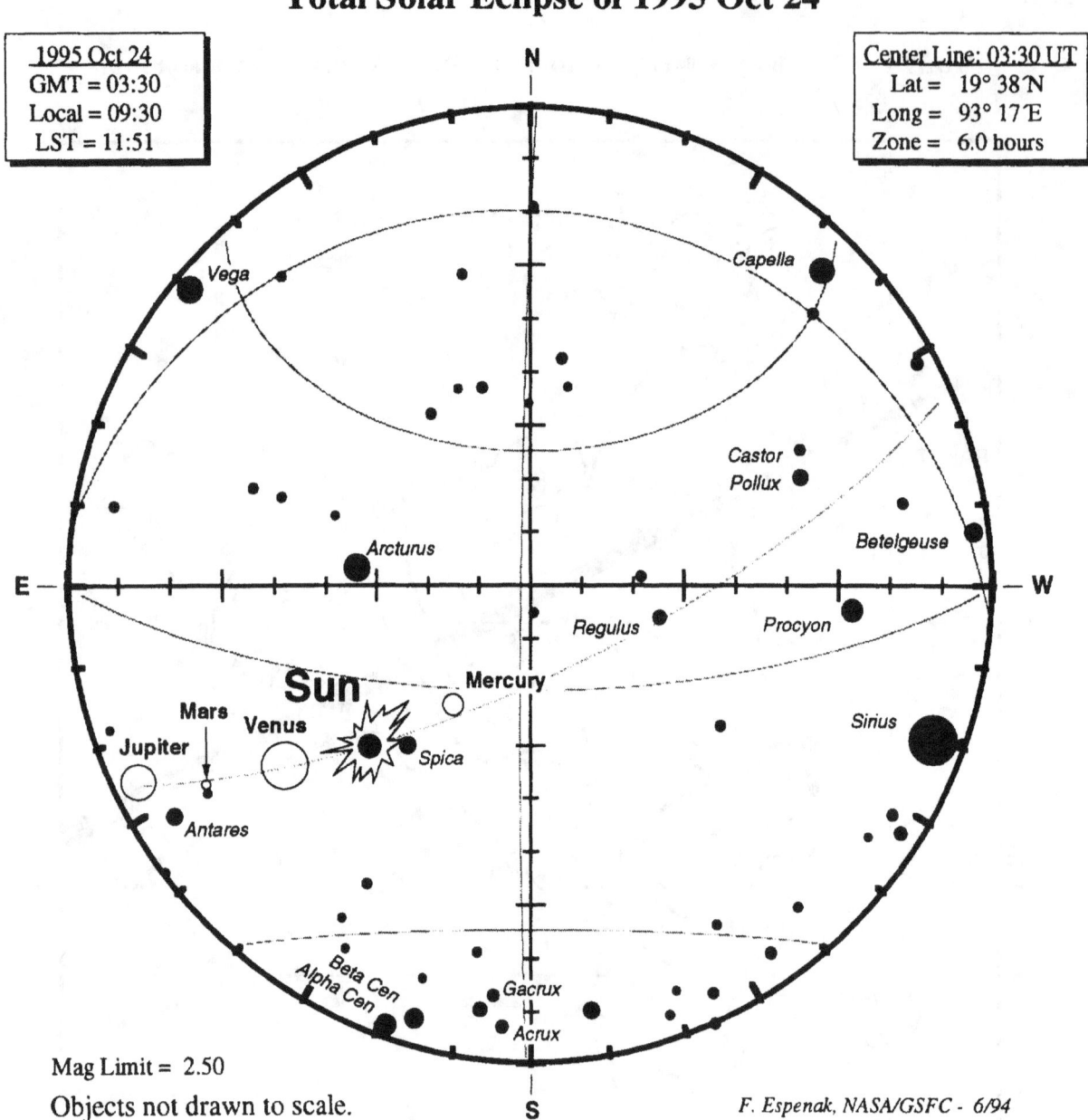

Figure 11: The sky during totality as seen from the center line in Myanmar at 03:30 UT. Jupiter (m=−1.4), Venus (m=−3.3) and Mercury (m=−0.6) should all be easily visible during the total eclipse. Bright stars which may also be visible include Spica (m=+1.0v), Antares (m=+0.9v), Arcturus (m=−0.04), Vega (m=+0.03), Regulus (m=+1.35), Alpha Centauri (m=−0.01), Beta Centauri (m=+0.6v), Acrux (m=+1.33) and Gacrux (m=+1.63v). Magnitudes followed by a 'v' are variable.

TOTAL SOLAR ECLIPSE OF 1995 OCTOBER 24

TABLES

Table 1

ELEMENTS OF THE TOTAL SOLAR ECLIPSE OF 1995 OCTOBER 24

Geocentric Conjunction   04:23:32.10 TDT   J.D. = 2450014.683010
of Sun & Moon in R.A.:   (=04:22:31.10 UT)

Instant of   04:33:30.49 TDT   J.D. = 2450014.689936
Greatest Eclipse: (=04:32:29.49 UT)

Geocentric Coordinates of Sun & Moon at Greatest Eclipse (DE200/LE200):

	Sun		Moon	
R.A.	= 13h 52m 45.401s	R.A.	= 13h 53m 07.189s
Dec.	= -11°-34´-24.47"	Dec.	= -11°-14´-17.00"
Semi-Diameter	= 16´ 04.68"	Semi-Diameter	= 16´ 10.19"
Eq.Hor.Par.	= 8.84"	Eq.Hor.Par.	= 0° 59´ 20.39"
Δ R.A.	= 9.525s/h	Δ R.A.	= 140.742s/h
Δ Dec.	= -52.38"/h	Δ Dec.	= -561.53"/h

Lunar Radius k_1 = 0.2725076 (Penumbra)   Shift in Δb = 0.0"
Constants: k_2 = 0.2722810 (Umbra)   Lunar Position: Δl = 0.0"

Geocentric Libration: l = -4.1°   Brown Lun. No. = 1185
(Optical + Physical) b = -0.4°   Saros Series = 143 (22/72)
                      c = 22.0°   Ephemeris = (DE200/LE200)

Eclipse Magnitude = 1.02134   Gamma = 0.35176 ΔT = 61.0 s

Polynomial Besselian Elements for: 1995 Oct 24   05:00:00.0 TDT   ($=t_0$)

n	x	y	d	l_1	l_2	μ
0	0.3300331	0.2763473	-11.5805674	0.5447736	-0.0013769	258.926819
1	0.5430526	-0.1441103	-0.0141956	-0.0001149	-0.0001143	15.002741
2	0.0000181	0.0000525	0.0000024	-0.0000121	-0.0000120	-0.000002
3	-0.0000083	0.0000021	0.0000000	0.0000000	0.0000000	0.000000

Tan f_1 = 0.0047003   Tan f_2 = 0.0046769

At time 't_1' (decimal hours), each besselian element is evaluated by:

$$x = x_0 + x_1*t + x_2*t^2 + x_3*t^3 \quad (\text{or } x = \Sigma\ [x_n*t^n];\ n = 0 \text{ to } 3)$$

where: $t = t_1 - t_0$ (decimal hours) and t_0 = 5.000

Note that all times are expressed in Terrestrial Dynamical Time (TDT).

Saros Series 143: Member 22 of 72 eclipses in series.

Table 2

SHADOW CONTACTS AND CIRCUMSTANCES
TOTAL SOLAR ECLIPSE OF 1995 OCTOBER 24

ΔT = 61.0 s
= 000°15.3´E

		Terrestrial Dynamical Time h m s	Latitude	Ephemeris Longitude†	True Longitude*
External/Internal Contacts of Penumbra:	P1	01:52:54.3	27°35.0´N	063°58.6´E	064°13.9´E
	P2	04:03:07.3	63°32.4´N	049°34.7´E	049°50.0´E
	P3	05:04:10.6	35°14.1´N	178°17.6´W	178°02.4´W
	P4	07:14:06.2	01°39.8´S	157°53.0´E	158°08.3´E
Extreme North/South Limits of Penumbral Path:	N1	03:44:13.9	74°03.7´N	075°46.6´E	076°01.9´E
	S1	02:53:46.9	03°48.3´N	043°25.7´E	043°41.0´E
	N2	05:23:11.7	47°49.6´N	172°11.6´E	172°26.9´E
	S2	06:13:07.2	25°26.1´S	178°23.6´E	178°38.9´E
External/Internal Contacts of Umbra:	U1	02:53:31.6	34°49.0´N	050°52.1´E	051°07.4´E
	U2	02:53:47.9	34°52.0´N	050°49.0´E	051°04.2´E
	U3	06:13:17.3	05°41.4´N	171°34.6´E	171°49.9´E
	U4	06:13:38.3	05°37.5´N	171°30.1´E	171°45.4´E
Extreme North/South Limits of Umbral Path:	N1	02:53:41.9	34°54.7´N	050°51.3´E	051°06.6´E
	S1	02:53:37.6	34°46.2´N	050°49.8´E	051°05.1´E
	N2	06:13:25.0	05°44.9´N	171°31.9´E	171°47.2´E
	S2	06:13:30.6	05°33.9´N	171°32.8´E	171°48.1´E
Extreme Limits of Center Line:	C1	02:53:39.7	34°50.5´N	050°50.5´E	051°05.8´E
	C2	06:13:27.8	05°39.4´N	171°32.3´E	171°47.6´E
Instant of Greatest Eclipse:	G0	04:33:30.5	08°24.6´N	112°55.9´E	113°11.2´E

Circumstances at Greatest Eclipse: Sun's Altitude = 69.4°   Path Width = 77.6 km
Sun's Azimuth = 194.0°   Central Duration = 02m09.2s

---

† Ephemeris Longitude is the terrestrial dynamical longitude assuming a uniformly rotating Earth.

* True Longitude is calculated by correcting the Ephemeris Longitude for the non-uniform rotation of Earth.
(T.L. = E.L. - 1.002738*ΔT/240, where ΔT (in seconds) = TDT - UT)

Note: Longitude is measured positive to the East.

Since ΔT is not known in advance, the value used in the predictions is an extrapolation based on pre-1993 measurements. Nevertheless, the actual value is expected to fall within ±0.3 seconds of the estimated ΔT used here.

Table 3

PATH OF THE UMBRAL SHADOW
TOTAL SOLAR ECLIPSE OF 1995 OCTOBER 24

Universal Time	Northern Limit Latitude	Northern Limit Longitude	Southern Limit Latitude	Southern Limit Longitude	Center Line Latitude	Center Line Longitude	Path Azm °	Path Width km	Central Durat.
Limits	34°54.7′N	051°06.6′E	34°46.2′N	051°05.1′E	34°50.5′N	051°05.8′E	–	16	00m16s
02:55	31°50.7′N	063°49.0′E	31°30.8′N	064°02.3′E	31°40.7′N	063°55.8′E	109	28	00m31s
03:00	29°04.7′N	072°35.9′E	28°41.7′N	072°39.4′E	28°53.2′N	072°37.8′E	112	37	00m44s
03:05	27°02.9′N	078°03.3′E	26°37.6′N	078°02.4′E	26°50.2′N	078°03.0′E	114	43	00m55s
03:10	25°19.8′N	082°12.2′E	24°52.7′N	082°08.2′E	25°06.2′N	082°10.3′E	116	48	01m03s
03:15	23°48.1′N	085°36.2′E	23°19.5′N	085°29.8′E	23°33.8′N	085°33.1′E	117	52	01m11s
03:20	22°24.4′N	088°30.9′E	21°54.6′N	088°22.5′E	22°09.4′N	088°26.8′E	118	55	01m18s
03:25	21°06.8′N	091°04.6′E	20°35.9′N	090°54.6′E	20°51.3′N	090°59.7′E	119	58	01m24s
03:30	19°54.1′N	093°22.7′E	19°22.3′N	093°11.3′E	19°38.2′N	093°17.1′E	120	61	01m30s
03:35	18°45.6′N	095°28.6′E	18°12.9′N	095°16.1′E	18°29.2′N	095°22.4′E	120	63	01m36s
03:40	17°40.6′N	097°25.0′E	17°07.2′N	097°11.4′E	17°23.8′N	097°18.3′E	121	65	01m41s
03:45	16°38.7′N	099°13.7′E	16°04.6′N	098°59.3′E	16°21.6′N	099°06.6′E	121	67	01m46s
03:50	15°39.5′N	100°56.2′E	15°04.8′N	100°41.0′E	15°22.1′N	100°48.7′E	121	69	01m50s
03:55	14°42.9′N	102°33.7′E	14°07.6′N	102°17.8′E	14°25.2′N	102°25.8′E	121	70	01m54s
04:00	13°48.6′N	104°07.0′E	13°12.8′N	103°50.6′E	13°30.7′N	103°58.8′E	121	72	01m57s
04:05	12°56.4′N	105°37.0′E	12°20.2′N	105°20.1′E	12°38.3′N	105°28.6′E	121	73	02m00s
04:10	12°06.3′N	107°04.5′E	11°29.6′N	106°47.2′E	11°47.9′N	106°55.9′E	120	74	02m03s
04:15	11°18.1′N	108°29.9′E	10°41.0′N	108°12.3′E	10°59.5′N	108°21.1′E	120	75	02m05s
04:20	10°31.7′N	109°53.8′E	09°54.3′N	109°36.0′E	10°13.0′N	109°44.9′E	119	76	02m07s
04:25	09°47.1′N	111°16.8′E	09°09.4′N	110°58.8′E	09°28.3′N	111°07.8′E	118	77	02m08s
04:30	09°04.3′N	112°39.2′E	08°26.3′N	112°21.2′E	08°45.3′N	112°30.2′E	117	77	02m09s
04:35	08°23.2′N	114°01.6′E	07°45.0′N	113°43.5′E	08°04.1′N	113°52.6′E	116	78	02m10s
04:40	07°43.7′N	115°24.3′E	07°05.4′N	115°06.3′E	07°24.5′N	115°15.3′E	115	78	02m10s
04:45	07°05.9′N	116°47.8′E	06°27.5′N	116°29.9′E	06°46.7′N	116°38.9′E	114	78	02m09s
04:50	06°29.9′N	118°12.6′E	05°51.4′N	117°54.9′E	06°10.7′N	118°03.7′E	112	78	02m08s
04:55	05°55.5′N	119°39.0′E	05°17.1′N	119°21.5′E	05°36.3′N	119°30.3′E	111	78	02m06s
05:00	05°23.0′N	121°07.7′E	04°44.6′N	120°50.5′E	05°03.8′N	120°59.1′E	109	77	02m04s
05:05	04°52.2′N	122°39.1′E	04°14.1′N	122°22.2′E	04°33.2′N	122°30.6′E	108	77	02m02s
05:10	04°23.4′N	124°13.9′E	03°45.5′N	123°57.3′E	04°04.5′N	124°05.6′E	106	76	01m59s
05:15	03°56.6′N	125°52.8′E	03°19.1′N	125°36.6′E	03°37.9′N	125°44.7′E	104	74	01m55s
05:20	03°32.0′N	127°36.6′E	02°54.9′N	127°20.7′E	03°13.5′N	127°28.6′E	102	73	01m51s
05:25	03°09.7′N	129°26.3′E	02°33.2′N	129°10.9′E	02°51.5′N	129°18.5′E	100	71	01m47s
05:30	02°50.1′N	131°23.2′E	02°14.2′N	131°08.2′E	02°32.2′N	131°15.6′E	98	69	01m42s
05:35	02°33.3′N	133°28.9′E	01°58.3′N	133°14.2′E	02°15.9′N	133°21.5′E	96	67	01m37s
05:40	02°20.0′N	135°45.4′E	01°46.0′N	135°31.2′E	02°03.0′N	135°38.2′E	94	64	01m31s
05:45	02°10.6′N	138°15.9′E	01°37.8′N	138°02.0′E	01°54.2′N	138°08.9′E	92	61	01m24s
05:50	02°06.2′N	141°04.7′E	01°34.8′N	140°51.0′E	01°50.5′N	140°57.8′E	90	58	01m17s
05:55	02°08.2′N	144°18.8′E	01°38.5′N	144°05.2′E	01°53.4′N	144°11.9′E	88	54	01m09s
06:00	02°19.3′N	148°10.8′E	01°51.7′N	147°56.9′E	02°05.6′N	148°03.8′E	86	49	01m00s
06:05	02°45.4′N	153°08.7′E	02°20.4′N	152°53.6′E	02°32.9′N	153°01.1′E	84	43	00m50s
06:10	03°46.4′N	160°49.8′E	03°25.0′N	160°29.5′E	03°35.7′N	160°39.5′E	81	33	00m36s
Limits	05°44.9′N	171°47.2′E	05°33.9′N	171°48.1′E	05°39.4′N	171°47.6′E	–	20	00m20s

Table 4

PHYSICAL EPHEMERIS OF THE UMBRAL SHADOW
TOTAL SOLAR ECLIPSE OF 1995 OCTOBER 24

Universal Time	Center Line Latitude	Center Line Longitude	Diameter Ratio	Eclipse Obscur.	Sun Alt °	Sun Azm °	Path Width km	Major Axis km	Minor Axis km	Umbra Veloc. km/s	Central Durat.
02:52.6	34°50.5´N	051°05.8´E	1.0044	1.0088	0.0	104.1	15.8	-	15.2	-	00m16s
02:55	31°40.7´N	063°55.8´E	1.0079	1.0159	11.6	111.6	28.1	134.7	27.3	4.216	00m31s
03:00	28°53.2´N	072°37.8´E	1.0105	1.0212	20.6	116.9	37.3	102.9	36.3	2.255	00m44s
03:05	26°50.2´N	078°03.0´E	1.0123	1.0247	26.8	120.4	43.3	93.7	42.3	1.677	00m55s
03:10	25°06.2´N	082°10.3´E	1.0136	1.0275	31.8	123.4	48.0	88.9	46.9	1.376	01m03s
03:15	23°33.8´N	085°33.1´E	1.0148	1.0298	36.1	126.1	51.9	85.9	50.7	1.186	01m11s
03:20	22°09.4´N	088°26.8´E	1.0157	1.0317	40.0	128.6	55.2	83.9	54.0	1.053	01m18s
03:25	20°51.3´N	090°59.7´E	1.0166	1.0334	43.6	131.2	58.2	82.4	56.8	0.954	01m24s
03:30	19°38.2´N	093°17.1´E	1.0173	1.0349	46.8	133.7	60.8	81.2	59.3	0.877	01m30s
03:35	18°29.2´N	095°22.4´E	1.0180	1.0363	49.9	136.4	63.1	80.3	61.5	0.816	01m36s
03:40	17°23.8´N	097°18.3´E	1.0185	1.0374	52.7	139.3	65.3	79.6	63.4	0.767	01m41s
03:45	16°21.6´N	099°06.6´E	1.0191	1.0385	55.4	142.4	67.2	79.1	65.1	0.726	01m46s
03:50	15°22.1´N	100°48.7´E	1.0195	1.0394	57.8	145.8	68.9	78.7	66.6	0.692	01m50s
03:55	14°25.2´N	102°25.8´E	1.0199	1.0402	60.1	149.6	70.5	78.3	67.9	0.663	01m54s
04:00	13°30.7´N	103°58.8´E	1.0202	1.0409	62.2	153.9	71.9	78.1	69.1	0.639	01m57s
04:05	12°38.3´N	105°28.6´E	1.0205	1.0415	64.1	158.7	73.1	77.9	70.1	0.620	02m00s
04:10	11°47.9´N	106°55.9´E	1.0208	1.0420	65.7	164.0	74.2	77.7	70.9	0.603	02m03s
04:15	10°59.5´N	108°21.1´E	1.0210	1.0424	67.1	170.0	75.2	77.6	71.6	0.590	02m05s
04:20	10°13.0´N	109°44.9´E	1.0211	1.0427	68.2	176.5	76.1	77.6	72.1	0.579	02m07s
04:25	09°28.3´N	111°07.8´E	1.0212	1.0429	68.9	183.6	76.8	77.6	72.5	0.571	02m08s
04:30	08°45.3´N	112°30.2´E	1.0213	1.0431	69.3	190.9	77.3	77.7	72.7	0.565	02m09s
04:35	08°04.1´N	113°52.6´E	1.0214	1.0432	69.3	198.4	77.7	77.8	72.8	0.563	02m10s
04:40	07°24.5´N	115°15.3´E	1.0213	1.0431	68.9	205.8	78.0	78.0	72.8	0.562	02m10s
04:45	06°46.7´N	116°38.9´E	1.0213	1.0430	68.2	212.7	78.1	78.2	72.6	0.564	02m09s
04:50	06°10.7´N	118°03.7´E	1.0212	1.0429	67.1	219.1	78.0	78.5	72.3	0.569	02m08s
04:55	05°36.3´N	119°30.3´E	1.0211	1.0426	65.8	224.8	77.7	78.8	71.9	0.577	02m06s
05:00	05°03.8´N	120°59.1´E	1.0209	1.0422	64.1	229.9	77.2	79.2	71.3	0.588	02m04s
05:05	04°33.2´N	122°30.6´E	1.0207	1.0418	62.3	234.3	76.5	79.7	70.5	0.602	02m02s
05:10	04°04.5´N	124°05.6´E	1.0204	1.0412	60.2	238.1	75.6	80.2	69.6	0.621	01m59s
05:15	03°37.9´N	125°44.7´E	1.0201	1.0405	57.9	241.3	74.4	80.9	68.5	0.645	01m55s
05:20	03°13.5´N	127°28.6´E	1.0197	1.0397	55.4	244.1	73.0	81.6	67.2	0.674	01m51s
05:25	02°51.5´N	129°18.5´E	1.0192	1.0388	52.8	246.5	71.3	82.5	65.7	0.711	01m47s
05:30	02°32.2´N	131°15.6´E	1.0187	1.0378	50.0	248.6	69.4	83.6	64.0	0.757	01m42s
05:35	02°15.9´N	133°21.5´E	1.0181	1.0366	46.9	250.3	67.1	85.0	62.1	0.814	01m37s
05:40	02°03.0´N	135°38.2´E	1.0175	1.0353	43.7	251.8	64.4	86.6	59.8	0.887	01m31s
05:45	01°54.2´N	138°08.9´E	1.0167	1.0337	40.1	253.1	61.4	88.8	57.2	0.983	01m24s
05:50	01°50.5´N	140°57.8´E	1.0158	1.0318	36.2	254.2	57.9	91.7	54.2	1.113	01m17s
05:55	01°53.4´N	144°11.9´E	1.0147	1.0297	31.9	255.1	53.8	95.7	50.6	1.300	01m09s
06:00	02°05.6´N	148°03.8´E	1.0134	1.0271	26.9	255.9	48.9	102.1	46.2	1.597	01m00s
06:05	02°32.9´N	153°01.1´E	1.0118	1.0237	20.8	256.6	42.6	114.2	40.5	2.167	00m50s
06:10	03°35.7´N	160°39.5´E	1.0092	1.0185	11.9	257.3	33.2	154.4	31.8	4.067	00m36s
06:12.4	05°39.4´N	171°47.6´E	1.0057	1.0114	0.0	258.3	20.4	-	19.6	-	00m20s

Table 5

LOCAL CIRCUMSTANCES ON THE CENTER LINE
TOTAL SOLAR ECLIPSE OF 1995 OCTOBER 24

Center Line Maximum Eclipse			First Contact				Second Contact			Third Contact			Fourth Contact			
U.T.	Durat.	Alt °	U.T.	P °	V °	Alt °	U.T.	P °	V °	U.T.	P °	V °	U.T.	P °	V °	Alt °
02:55	00m31s	12	-	-	-	-	02:54:45	112	166	02:55:16	292	346	04:04:36	113	160	25
03:00	00m44s	21	01:53:17	292	351	7	02:59:38	113	167	03:00:22	293	346	04:15:18	114	157	34
03:05	00m55s	27	01:55:02	293	352	13	03:04:33	114	166	03:05:27	294	346	04:24:26	115	154	41
03:10	01m03s	32	01:57:14	294	353	17	03:09:29	115	166	03:10:32	295	346	04:32:52	116	150	46
03:15	01m11s	36	01:59:43	295	354	21	03:14:25	116	165	03:15:36	296	345	04:40:50	117	146	50
03:20	01m18s	40	02:02:23	295	354	25	03:19:21	116	164	03:20:39	296	344	04:48:27	117	141	53
03:25	01m24s	44	02:05:11	296	355	28	03:24:18	117	163	03:25:42	297	343	04:55:45	117	136	56
03:30	01m30s	47	02:08:07	296	355	31	03:29:15	117	161	03:30:45	297	341	05:02:47	117	130	58
03:35	01m36s	50	02:11:10	297	355	34	03:34:12	117	160	03:35:48	297	339	05:09:33	117	124	60
03:40	01m41s	53	02:14:18	297	355	37	03:39:10	118	157	03:40:51	298	337	05:16:06	117	117	61
03:45	01m46s	55	02:17:32	297	355	39	03:44:07	118	155	03:45:53	298	334	05:22:24	117	110	62
03:50	01m50s	58	02:20:50	298	354	42	03:49:05	118	152	03:50:55	298	331	05:28:30	117	103	62
03:55	01m54s	60	02:24:14	298	354	45	03:54:03	118	148	03:55:57	298	328	05:34:23	117	96	62
04:00	01m57s	62	02:27:43	298	353	47	03:59:01	118	144	04:00:59	298	323	05:40:03	116	89	62
04:05	02m00s	64	02:31:18	298	352	49	04:04:00	118	140	04:06:00	298	318	05:45:32	116	83	61
04:10	02m03s	66	02:34:58	298	351	52	04:08:59	118	134	04:11:02	298	313	05:50:49	116	77	60
04:15	02m05s	67	02:38:43	298	350	54	04:13:57	117	128	04:16:03	297	307	05:55:55	115	71	59
04:20	02m07s	68	02:42:35	298	349	56	04:18:57	117	121	04:21:04	297	300	06:00:50	115	66	58
04:25	02m08s	69	02:46:33	298	347	59	04:23:56	117	114	04:26:04	297	292	06:05:35	114	62	56
04:30	02m09s	69	02:50:37	298	344	61	04:28:55	116	106	04:31:05	296	284	06:10:10	114	58	55
04:35	02m10s	69	02:54:49	298	342	63	04:33:55	116	98	04:36:05	296	276	06:14:36	113	54	53
04:40	02m10s	69	02:59:09	297	338	65	04:38:55	115	90	04:41:05	295	269	06:18:53	112	50	51
04:45	02m09s	68	03:03:37	297	334	67	04:43:56	115	82	04:46:04	295	261	06:23:01	112	47	49
04:50	02m08s	67	03:08:14	297	329	69	04:48:56	114	75	04:51:04	294	254	06:27:01	111	44	48
04:55	02m06s	66	03:13:01	296	322	71	04:53:57	114	68	04:56:03	294	247	06:30:52	111	41	46
05:00	02m04s	64	03:17:59	296	315	72	04:58:58	113	62	05:01:02	293	242	06:34:36	110	38	44
05:05	02m02s	62	03:23:07	295	305	74	05:03:59	112	57	05:06:01	292	236	06:38:13	109	36	41
05:10	01m59s	60	03:28:29	295	294	74	05:09:00	112	52	05:10:59	292	231	06:41:43	109	34	39
05:15	01m55s	58	03:34:03	294	282	74	05:14:02	111	48	05:15:58	291	227	06:45:06	108	32	37
05:20	01m51s	55	03:39:52	293	270	74	05:19:04	110	44	05:20:56	290	223	06:48:23	107	29	35
05:25	01m47s	53	03:45:56	292	258	73	05:24:06	109	40	05:25:53	289	220	06:51:33	107	27	32
05:30	01m42s	50	03:52:17	292	248	71	05:29:09	108	37	05:30:51	288	217	06:54:36	106	26	30
05:35	01m37s	47	03:58:56	291	239	68	05:34:12	108	34	05:35:48	288	214	06:57:32	105	24	27
05:40	01m31s	44	04:05:55	290	231	65	05:39:15	107	31	05:40:45	287	211	07:00:21	104	22	24
05:45	01m24s	40	04:13:15	289	225	61	05:44:18	106	28	05:45:42	286	208	07:03:01	104	20	21
05:50	01m17s	36	04:20:58	288	220	57	05:49:21	105	26	05:50:38	285	206	07:05:32	103	19	18
05:55	01m09s	32	04:29:08	286	215	52	05:54:25	104	24	05:55:35	284	204	07:07:51	102	17	14
06:00	01m00s	27	04:37:52	285	211	47	05:59:30	103	22	06:00:30	283	201	07:09:54	102	16	10
06:05	00m50s	21	04:47:25	284	207	39	06:04:35	102	19	06:05:25	282	199	07:11:28	101	14	5
06:10	00m36s	12	04:58:41	282	203	29	06:09:42	101	17	06:10:18	281	197	-	-	-	-

Table 6

TOPOCENTRIC DATA AND PATH CORRECTIONS DUE TO LUNAR LIMB PROFILE

Universal Time	Moon Topo H.P. "	Moon Topo S.D. "	Moon Rel. Ang.V "/s	Topo Lib. Long °	Sun Alt. °	Sun Az. °	Path Az. °	North Limit P.A. °	North Limit		South Limit		Central Durat. Cor. s
									Int. '	Ext. '	Int. '	Ext. '	
02:55	3570.6	972.3	0.491	-3.22	11.6	111.6	108.7	22.0	-0.3	0.6	1.1	-1.9	-6.5
03:00	3580.0	974.8	0.457	-3.26	20.6	116.9	112.1	23.5	-0.3	0.6	0.4	-2.3	-6.8
03:05	3586.2	976.5	0.435	-3.30	26.8	120.4	114.2	24.5	-0.2	0.7	0.4	-2.7	-7.5
03:10	3591.1	977.9	0.417	-3.35	31.8	123.4	115.9	25.2	-0.2	0.6	0.4	-2.8	-7.7
03:15	3595.2	978.9	0.403	-3.39	36.1	126.1	117.2	25.9	-0.2	0.5	0.4	-2.7	-8.0
03:20	3598.6	979.9	0.390	-3.43	40.0	128.6	118.3	26.4	-0.2	0.6	0.4	-2.6	-8.2
03:25	3601.6	980.7	0.380	-3.47	43.6	131.2	119.2	26.8	-0.1	0.7	0.3	-2.5	-8.4
03:30	3604.3	981.4	0.370	-3.52	46.8	133.7	119.9	27.2	-0.1	0.8	0.3	-2.3	-8.4
03:35	3606.6	982.0	0.362	-3.56	49.9	136.4	120.5	27.4	-0.1	0.8	0.3	-2.3	-8.6
03:40	3608.7	982.6	0.354	-3.60	52.7	139.3	120.9	27.7	-0.0	0.9	0.3	-2.4	-8.7
03:45	3610.6	983.1	0.348	-3.64	55.4	142.4	121.1	27.8	-0.0	0.9	0.3	-2.5	-8.9
03:50	3612.2	983.5	0.342	-3.69	57.8	145.8	121.2	27.9	-0.0	0.9	0.3	-2.6	-9.0
03:55	3613.6	983.9	0.337	-3.73	60.1	149.6	121.2	27.9	0.0	0.9	0.3	-2.6	-9.2
04:00	3614.8	984.2	0.333	-3.77	62.2	153.9	121.1	27.8	0.0	0.9	0.3	-2.6	-9.3
04:05	3615.9	984.5	0.329	-3.81	64.1	158.7	120.8	27.7	0.0	0.8	0.3	-2.5	-9.4
04:10	3616.8	984.8	0.326	-3.86	65.7	164.0	120.3	27.5	0.0	0.8	0.3	-2.4	-9.4
04:15	3617.5	984.9	0.323	-3.90	67.1	170.0	119.8	27.3	-0.0	0.8	0.3	-2.3	-9.5
04:20	3618.1	985.1	0.321	-3.94	68.2	176.5	119.1	27.0	-0.0	0.8	0.3	-2.4	-9.5
04:25	3618.5	985.2	0.319	-3.98	68.9	183.6	118.3	26.7	-0.1	0.8	0.4	-2.6	-9.5
04:30	3618.8	985.3	0.318	-4.03	69.3	190.9	117.4	26.3	-0.1	0.7	0.4	-2.7	-9.5
04:35	3618.9	985.3	0.318	-4.07	69.3	198.4	116.3	25.9	-0.1	0.5	0.4	-2.9	-9.4
04:40	3618.9	985.3	0.318	-4.11	68.9	205.8	115.1	25.4	-0.1	0.7	0.4	-2.9	-8.9
04:45	3618.7	985.3	0.319	-4.15	68.2	212.7	113.9	24.9	-0.1	0.8	0.5	-2.9	-8.7
04:50	3618.4	985.2	0.320	-4.20	67.1	219.1	112.5	24.3	-0.2	0.9	0.5	-2.8	-8.6
04:55	3617.9	985.0	0.322	-4.24	65.8	224.8	111.0	23.7	-0.2	0.8	0.5	-2.6	-8.4
05:00	3617.2	984.9	0.324	-4.28	64.1	229.9	109.4	23.0	-0.2	0.6	0.5	-2.2	-8.1
05:05	3616.4	984.7	0.327	-4.32	62.3	234.3	107.7	22.3	-0.2	0.6	1.2	-1.9	-8.0
05:10	3615.4	984.4	0.331	-4.37	60.2	238.1	106.0	21.6	-0.2	0.8	1.2	-1.9	-7.7
05:15	3614.3	984.1	0.335	-4.41	57.9	241.3	104.2	20.9	-0.2	0.7	1.2	-1.7	-7.4
05:20	3612.9	983.7	0.341	-4.45	55.4	244.1	102.3	20.1	-0.2	0.5	1.3	-1.4	-7.0
05:25	3611.3	983.3	0.347	-4.49	52.8	246.5	100.4	19.3	-0.2	0.6	1.3	-1.2	-6.7
05:30	3609.5	982.8	0.354	-4.54	50.0	248.6	98.4	18.5	-0.2	0.5	1.3	-1.4	-6.3
05:35	3607.4	982.2	0.362	-4.58	46.9	250.3	96.4	17.6	-0.2	0.5	1.4	-1.5	-6.0
05:40	3604.9	981.6	0.372	-4.62	43.7	251.8	94.4	16.7	-0.1	0.6	1.4	-1.5	-5.7
05:45	3602.2	980.8	0.383	-4.66	40.1	253.1	92.3	15.9	-0.1	0.7	1.4	-1.6	-5.4
05:50	3598.9	980.0	0.396	-4.71	36.2	254.2	90.2	15.0	-0.1	0.6	1.4	-1.3	-4.8
05:55	3595.1	978.9	0.411	-4.75	31.9	255.1	88.1	14.0	-0.3	0.6	1.4	-0.9	-4.6
06:00	3590.5	977.7	0.429	-4.79	26.9	255.9	85.9	13.1	-0.4	0.8	1.4	-0.6	-4.5
06:05	3584.5	976.1	0.453	-4.83	20.8	256.6	83.6	12.1	-0.4	0.7	1.4	-1.5	-4.5
06:10	3575.4	973.6	0.489	-4.88	11.9	257.3	80.9	10.8	-0.4	0.7	1.4	-2.4	-4.9

Table 7
MAPPING COORDINATES FOR THE UMBRAL PATH

Longitude	Latitude of:			Universal Time at:			Circumstances on the Center Line			
	Northern Limit	Southern Limit	Center Line	Northern Limit	Southern Limit	Center Line	Sun Alt °	Sun Az. °	Path Width km	Center Durat.
				h m s	h m s	h m s				
052°00.0´E	34°43.7´N	34°34.5´N	34°39.1´N	02:52:42	02:52:38	02:52:39	1	105	17	00m17s
053°00.0´E	34°31.1´N	34°21.3´N	34°26.2´N	02:52:44	02:52:39	02:52:41	2	105	18	00m18s
054°00.0´E	34°18.1´N	34°07.7´N	34°12.9´N	02:52:48	02:52:43	02:52:45	2	106	18	00m19s
055°00.0´E	34°04.7´N	33°53.8´N	33°59.3´N	02:52:53	02:52:48	02:52:51	3	106	19	00m20s
056°00.0´E	33°51.0´N	33°39.6´N	33°45.3´N	02:53:00	02:52:55	02:52:58	4	107	20	00m21s
057°00.0´E	33°36.9´N	33°24.9´N	33°30.9´N	02:53:09	02:53:04	02:53:06	5	108	21	00m22s
058°00.0´E	33°22.5´N	33°09.9´N	33°16.2´N	02:53:20	02:53:14	02:53:17	6	108	22	00m23s
059°00.0´E	33°07.6´N	32°54.5´N	33°01.1´N	02:53:32	02:53:27	02:53:29	7	109	23	00m25s
060°00.0´E	32°52.4´N	32°38.7´N	32°45.6´N	02:53:47	02:53:41	02:53:44	8	109	24	00m26s
061°00.0´E	32°36.8´N	32°22.5´N	32°29.7´N	02:54:03	02:53:57	02:54:00	9	110	25	00m27s
062°00.0´E	32°20.8´N	32°05.9´N	32°13.4´N	02:54:21	02:54:16	02:54:19	10	110	26	00m28s
063°00.0´E	32°04.4´N	31°48.9´N	31°56.6´N	02:54:42	02:54:36	02:54:39	11	111	27	00m30s
064°00.0´E	31°47.6´N	31°31.5´N	31°39.5´N	02:55:04	02:54:59	02:55:02	12	112	28	00m31s
065°00.0´E	31°30.3´N	31°13.6´N	31°22.0´N	02:55:29	02:55:24	02:55:27	13	112	29	00m33s
066°00.0´E	31°12.6´N	30°55.3´N	31°04.0´N	02:55:56	02:55:51	02:55:54	14	113	30	00m34s
067°00.0´E	30°54.5´N	30°36.5´N	30°45.6´N	02:56:26	02:56:21	02:56:24	15	113	31	00m35s
068°00.0´E	30°36.0´N	30°17.3´N	30°26.7´N	02:56:58	02:56:54	02:56:56	16	114	32	00m37s
069°00.0´E	30°17.0´N	29°57.7´N	30°07.3´N	02:57:33	02:57:28	02:57:31	17	115	33	00m39s
070°00.0´E	29°57.5´N	29°37.6´N	29°47.5´N	02:58:10	02:58:06	02:58:08	18	115	34	00m40s
071°00.0´E	29°37.6´N	29°16.9´N	29°27.3´N	02:58:50	02:58:47	02:58:48	19	116	36	00m42s
072°00.0´E	29°17.2´N	28°55.8´N	29°06.5´N	02:59:33	02:59:30	02:59:31	20	116	37	00m43s
073°00.0´E	28°56.3´N	28°34.3´N	28°45.3´N	03:00:19	03:00:16	03:00:17	21	117	38	00m45s
074°00.0´E	28°34.9´N	28°12.2´N	28°23.5´N	03:01:08	03:01:06	03:01:07	22	118	39	00m47s
075°00.0´E	28°13.0´N	27°49.6´N	28°01.3´N	03:02:00	03:01:58	03:01:59	23	118	40	00m49s
076°00.0´E	27°50.6´N	27°26.4´N	27°38.5´N	03:02:55	03:02:55	03:02:55	24	119	41	00m51s
077°00.0´E	27°27.7´N	27°02.8´N	27°15.2´N	03:03:54	03:03:54	03:03:54	26	120	42	00m52s
078°00.0´E	27°04.2´N	26°38.6´N	26°51.4´N	03:04:56	03:04:57	03:04:57	27	120	43	00m54s
079°00.0´E	26°40.2´N	26°13.9´N	26°27.1´N	03:06:02	03:06:04	03:06:03	28	121	44	00m56s
080°00.0´E	26°15.7´N	25°48.6´N	26°02.2´N	03:07:12	03:07:15	03:07:14	29	122	45	00m59s
081°00.0´E	25°50.7´N	25°22.7´N	25°36.7´N	03:08:26	03:08:30	03:08:28	30	123	47	01m01s
082°00.0´E	25°25.0´N	24°56.3´N	25°10.7´N	03:09:44	03:09:49	03:09:46	32	123	48	01m03s
083°00.0´E	24°58.9´N	24°29.4´N	24°44.1´N	03:11:06	03:11:12	03:11:09	33	124	49	01m05s
084°00.0´E	24°32.1´N	24°01.8´N	24°17.0´N	03:12:32	03:12:40	03:12:36	34	125	50	01m07s
085°00.0´E	24°04.8´N	23°33.7´N	23°49.3´N	03:14:03	03:14:12	03:14:07	35	126	51	01m10s
086°00.0´E	23°36.9´N	23°05.0´N	23°21.0´N	03:15:38	03:15:49	03:15:44	37	126	52	01m12s
087°00.0´E	23°08.5´N	22°35.8´N	22°52.2´N	03:17:19	03:17:31	03:17:25	38	127	54	01m14s
088°00.0´E	22°39.5´N	22°05.9´N	22°22.8´N	03:19:04	03:19:18	03:19:11	39	128	55	01m17s
089°00.0´E	22°09.9´N	21°35.5´N	21°52.8´N	03:20:54	03:21:11	03:21:02	41	129	56	01m19s
090°00.0´E	21°39.8´N	21°04.6´N	21°22.2´N	03:22:50	03:23:08	03:22:59	42	130	57	01m22s
091°00.0´E	21°09.2´N	20°33.1´N	20°51.2´N	03:24:51	03:25:11	03:25:01	44	131	58	01m24s
092°00.0´E	20°38.0´N	20°01.0´N	20°19.5´N	03:26:57	03:27:20	03:27:08	45	132	59	01m27s
093°00.0´E	20°06.2´N	19°28.5´N	19°47.4´N	03:29:09	03:29:34	03:29:21	46	133	60	01m30s
094°00.0´E	19°34.0´N	18°55.5´N	19°14.8´N	03:31:26	03:31:54	03:31:40	48	135	62	01m32s
095°00.0´E	19°01.3´N	18°22.0´N	18°41.7´N	03:33:50	03:34:20	03:34:04	49	136	63	01m35s
096°00.0´E	18°28.2´N	17°48.1´N	18°08.2´N	03:36:19	03:36:52	03:36:35	51	137	64	01m37s
097°00.0´E	17°54.7´N	17°13.7´N	17°34.3´N	03:38:54	03:39:29	03:39:11	52	139	65	01m40s
098°00.0´E	17°20.8´N	16°39.1´N	16°60.0´N	03:41:34	03:42:13	03:41:53	54	140	66	01m43s
099°00.0´E	16°46.5´N	16°04.1´N	16°25.4´N	03:44:21	03:45:02	03:44:41	55	142	67	01m45s
100°00.0´E	16°12.0´N	15°28.9´N	15°50.5´N	03:47:13	03:47:57	03:47:35	57	144	68	01m48s

Table 7
MAPPING COORDINATES FOR THE UMBRAL PATH

Longitude	Latitude of:			Universal Time at:			Circumstances on the Center Line			
	Northern Limit	Southern Limit	Center Line	Northern Limit	Southern Limit	Center Line	Sun Alt °	Sun Az. °	Path Width km	Center Durat.
				h m s	h m s	h m s				
101°00.0´E	15°37.3´N	14°53.6´N	15°15.5´N	03:50:11	03:50:58	03:50:34	58	146	69	01m50s
102°00.0´E	15°02.5´N	14°18.1´N	14°40.4´N	03:53:15	03:54:04	03:53:39	60	149	70	01m53s
103°00.0´E	14°27.6´N	13°42.6´N	14°05.1´N	03:56:23	03:57:15	03:56:49	61	151	71	01m55s
104°00.0´E	13°52.7´N	13°07.2´N	13°30.0´N	03:59:37	04:00:31	04:00:04	62	154	72	01m57s
105°00.0´E	13°17.8´N	12°31.9´N	12°54.9´N	04:02:55	04:03:52	04:03:23	63	157	73	01m59s
106°00.0´E	12°43.2´N	11°56.9´N	12°20.1´N	04:06:18	04:07:16	04:06:47	65	160	74	02m01s
107°00.0´E	12°08.8´N	11°22.2´N	11°45.6´N	04:09:45	04:10:45	04:10:14	66	164	74	02m03s
108°00.0´E	11°34.8´N	10°48.0´N	11°11.4´N	04:13:14	04:14:16	04:13:45	67	168	75	02m05s
109°00.0´E	11°01.3´N	10°14.2´N	10°37.8´N	04:16:47	04:17:50	04:17:19	68	173	76	02m06s
110°00.0´E	10°28.4´N	09°41.2´N	10°04.8´N	04:20:22	04:21:27	04:20:54	68	178	76	02m07s
111°00.0´E	09°56.0´N	09°08.8´N	09°32.4´N	04:23:59	04:25:04	04:24:32	69	183	77	02m08s
112°00.0´E	09°24.5´N	08°37.3´N	09°00.9´N	04:27:37	04:28:43	04:28:10	69	188	77	02m09s
113°00.0´E	08°53.7´N	08°06.6´N	08°30.2´N	04:31:16	04:32:22	04:31:49	69	194	78	02m09s
114°00.0´E	08°23.9´N	07°36.9´N	08°00.4´N	04:34:54	04:36:00	04:35:27	69	199	78	02m10s
115°00.0´E	07°55.1´N	07°08.3´N	07°31.7´N	04:38:32	04:39:37	04:39:05	69	204	78	02m10s
116°00.0´E	07°27.3´N	06°40.8´N	07°04.1´N	04:42:09	04:43:13	04:42:41	69	210	78	02m09s
117°00.0´E	07°00.6´N	06°14.4´N	06°37.5´N	04:45:43	04:46:47	04:46:15	68	214	78	02m09s
118°00.0´E	06°35.1´N	05°49.3´N	06°12.2´N	04:49:16	04:50:18	04:49:47	67	219	78	02m08s
119°00.0´E	06°10.7´N	05°25.4´N	05°48.0´N	04:52:45	04:53:46	04:53:16	66	223	78	02m07s
120°00.0´E	05°47.6´N	05°02.7´N	05°25.1´N	04:56:12	04:57:11	04:56:41	65	227	78	02m06s
121°00.0´E	05°25.7´N	04°41.3´N	05°03.5´N	04:59:34	05:00:32	05:00:03	64	230	77	02m04s
122°00.0´E	05°05.0´N	04°21.2´N	04°43.1´N	05:02:53	05:03:48	05:03:21	63	233	77	02m03s
123°00.0´E	04°45.6´N	04°02.3´N	04°24.0´N	05:06:07	05:07:01	05:06:34	62	236	76	02m01s
124°00.0´E	04°27.5´N	03°44.7´N	04°06.1´N	05:09:17	05:10:08	05:09:43	60	238	76	01m59s
125°00.0´E	04°10.5´N	03°28.4´N	03°49.4´N	05:12:21	05:13:11	05:12:46	59	240	75	01m57s
126°00.0´E	03°54.8´N	03°13.3´N	03°34.0´N	05:15:21	05:16:09	05:15:45	58	242	74	01m55s
127°00.0´E	03°40.3´N	02°59.4´N	03°19.8´N	05:18:16	05:19:02	05:18:39	56	243	73	01m53s
128°00.0´E	03°26.9´N	02°46.7´N	03°06.8´N	05:21:05	05:21:49	05:21:27	55	245	73	01m50s
129°00.0´E	03°14.7´N	02°35.2´N	02°54.9´N	05:23:50	05:24:31	05:24:11	53	246	72	01m48s
130°00.0´E	03°03.6´N	02°24.7´N	02°44.2´N	05:26:29	05:27:08	05:26:49	52	247	71	01m45s
131°00.0´E	02°53.6´N	02°15.4´N	02°34.5´N	05:29:02	05:29:40	05:29:21	50	248	70	01m43s
132°00.0´E	02°44.7´N	02°07.1´N	02°25.9´N	05:31:30	05:32:06	05:31:48	49	249	69	01m40s
133°00.0´E	02°36.8´N	01°59.9´N	02°18.3´N	05:33:53	05:34:27	05:34:10	47	250	67	01m38s
134°00.0´E	02°29.9´N	01°53.7´N	02°11.7´N	05:36:11	05:36:43	05:36:27	46	251	66	01m35s
135°00.0´E	02°23.9´N	01°48.4´N	02°06.1´N	05:38:23	05:38:54	05:38:39	45	251	65	01m32s
136°00.0´E	02°18.8´N	01°44.0´N	02°01.4´N	05:40:30	05:41:00	05:40:45	43	252	64	01m30s
137°00.0´E	02°14.7´N	01°40.5´N	01°57.6´N	05:42:33	05:43:00	05:42:47	42	253	63	01m27s
138°00.0´E	02°11.3´N	01°37.9´N	01°54.6´N	05:44:30	05:44:56	05:44:43	40	253	62	01m25s
139°00.0´E	02°08.9´N	01°36.1´N	01°52.5´N	05:46:22	05:46:47	05:46:35	39	253	60	01m22s
140°00.0´E	02°07.2´N	01°35.1´N	01°51.1´N	05:48:09	05:48:33	05:48:22	38	254	59	01m19s

Table 7
MAPPING COORDINATES FOR THE UMBRAL PATH

Longitude	Latitude of:			Universal Time at:			Circumstances on the Center Line			
	Northern Limit	Southern Limit	Center Line	Northern Limit	Southern Limit	Center Line	Sun Alt °	Sun Az. °	Path Width km	Center Durat.
				h m s	h m s	h m s				
141°00.0´E	02°06.2´N	01°34.8´N	01°50.5´N	05:49:52	05:50:15	05:50:04	36	254	58	01m17s
142°00.0´E	02°06.0´N	01°35.3´N	01°50.7´N	05:51:30	05:51:52	05:51:41	35	254	57	01m15s
143°00.0´E	02°06.5´N	01°36.5´N	01°51.5´N	05:53:04	05:53:24	05:53:14	33	255	55	01m12s
144°00.0´E	02°07.7´N	01°38.4´N	01°53.0´N	05:54:33	05:54:53	05:54:43	32	255	54	01m10s
145°00.0´E	02°09.5´N	01°40.9´N	01°55.2´N	05:55:58	05:56:17	05:56:07	31	255	53	01m07s
146°00.0´E	02°12.0´N	01°44.0´N	01°58.0´N	05:57:18	05:57:36	05:57:27	30	255	51	01m05s
147°00.0´E	02°15.0´N	01°47.7´N	02°01.3´N	05:58:35	05:58:52	05:58:44	28	256	50	01m03s
148°00.0´E	02°18.6´N	01°52.0´N	02°05.3´N	05:59:47	06:00:04	05:59:56	27	256	49	01m01s
149°00.0´E	02°22.8´N	01°56.8´N	02°09.8´N	06:00:56	06:01:11	06:01:04	26	256	48	00m58s
150°00.0´E	02°27.4´N	02°02.2´N	02°14.8´N	06:02:01	06:02:15	06:02:08	24	256	46	00m56s
151°00.0´E	02°32.6´N	02°08.0´N	02°20.3´N	06:03:01	06:03:16	06:03:09	23	256	45	00m54s
152°00.0´E	02°38.3´N	02°14.4´N	02°26.3´N	06:03:59	06:04:12	06:04:06	22	256	44	00m52s
153°00.0´E	02°44.4´N	02°21.2´N	02°32.8´N	06:04:53	06:05:05	06:04:59	21	257	43	00m50s
154°00.0´E	02°51.0´N	02°28.4´N	02°39.7´N	06:05:43	06:05:55	06:05:49	20	257	41	00m48s
155°00.0´E	02°58.0´N	02°36.1´N	02°47.0´N	06:06:30	06:06:42	06:06:36	18	257	40	00m46s
156°00.0´E	03°05.4´N	02°44.1´N	02°54.8´N	06:07:13	06:07:25	06:07:19	17	257	39	00m44s
157°00.0´E	03°13.2´N	02°52.6´N	03°02.9´N	06:07:54	06:08:05	06:07:59	16	257	38	00m43s
158°00.0´E	03°21.4´N	03°01.4´N	03°11.4´N	06:08:31	06:08:41	06:08:36	15	257	36	00m41s
159°00.0´E	03°29.9´N	03°10.6´N	03°20.3´N	06:09:05	06:09:15	06:09:10	14	257	35	00m39s
160°00.0´E	03°38.8´N	03°20.1´N	03°29.5´N	06:09:36	06:09:46	06:09:41	13	257	34	00m37s
161°00.0´E	03°48.0´N	03°30.0´N	03°39.0´N	06:10:05	06:10:14	06:10:09	11	257	33	00m36s
162°00.0´E	03°57.5´N	03°40.2´N	03°48.9´N	06:10:30	06:10:39	06:10:35	10	257	32	00m34s
163°00.0´E	04°07.4´N	03°50.6´N	03°59.0´N	06:10:53	06:11:01	06:10:57	9	258	30	00m33s
164°00.0´E	04°17.5´N	04°01.4´N	04°09.5´N	06:11:13	06:11:21	06:11:17	8	258	29	00m31s
165°00.0´E	04°27.9´N	04°12.4´N	04°20.2´N	06:11:30	06:11:38	06:11:34	7	258	28	00m30s
166°00.0´E	04°38.6´N	04°23.7´N	04°31.2´N	06:11:45	06:11:53	06:11:49	6	258	27	00m28s
167°00.0´E	04°49.5´N	04°35.3´N	04°42.4´N	06:11:57	06:12:05	06:12:01	5	258	26	00m27s
168°00.0´E	05°00.6´N	04°47.1´N	04°53.9´N	06:12:07	06:12:14	06:12:11	4	258	25	00m25s
169°00.0´E	05°12.0´N	04°59.1´N	05°05.6´N	06:12:15	06:12:21	06:12:18	3	258	23	00m24s
170°00.0´E	05°23.7´N	05°11.3´N	05°17.5´N	06:12:20	06:12:26	06:12:23	2	258	22	00m23s

Table 8a
CIRCUMSTANCES AT MAXIMUM ECLIPSE ON 1995 OCTOBER 24
FOR AFRICA, MIDDLE EAST AND RUSSIA

Location Name	Latitude	Longitude	Elev. m	U.T. of Maximum Eclipse h m s	Umbral Duration	Path Width km	Sun Alt °	Sun Azm °	P °	V °	Eclipse Mag.	Eclipse Obs.
ARMENIA												
Yerevan	40°10.0'N	044°30.0'E	–	03:22 Rise			0	105	–	–	0.524	0.419
AZERBAIJAN												
Baku	40°23.0'N	049°51.0'E	–	03:01 Rise			0	105	–	–	0.818	0.770
DJIBOUTI												
Djibouti	11°36.0'N	043°09.0'E	8	02:58 Rise			0	102	–	–	0.220	0.120
GEORGIA												
Tbilisi	41°40.0'N	044°45.0'E	–	03:23 Rise			0	105	–	–	0.512	0.404
IRAN												
Ahvaz	31°19.0'N	048°42.0'E	–	02:55 Rise			0	103	–	–	0.859	0.822
Birjand	32°53.0'N	059°13.0'E	–	02:53:30.5	00m17.4s	23	7	109	21	76	1.007	1.000
Esfahan	32°40.0'N	051°38.0'E	1719	02:51:40.6			1	104	20	77	0.940	0.926
Mashhad	36°18.0'N	059°36.0'E	–	02:55:12.3			7	110	201	252	0.902	0.878
Qom	34°39.0'N	050°54.0'E	–	02:52:31.8			0	104	20	75	0.995	0.995
Shiraz	29°36.0'N	052°32.0'E	–	02:50:43.0			2	105	21	80	0.853	0.815
Tabriz	38°05.0'N	046°18.0'E	–	03:12 Rise			0	104	–	–	0.678	0.598
Tehran	35°40.0'N	051°26.0'E	1292	02:53:05.1			1	104	200	254	0.975	0.970
IRAQ												
Baghdad	33°21.0'N	044°25.0'E	36	03:13 Rise			0	103	–	–	0.617	0.525
Basra	30°30.0'N	047°47.0'E	3	02:57 Rise			0	103	–	–	0.806	0.756
Mosul	36°20.0'N	043°08.0'E	240	03:20 Rise			0	104	–	–	0.534	0.429
KUWAIT												
Kuwait	29°20.0'N	047°59.0'E	5	02:55 Rise			0	103	–	–	0.793	0.739
QATAR												
Doha	25°17.0'N	051°32.0'E	–	02:49:33.9			2	104	21	84	0.714	0.642
RUSSIA												
Barnaul	53°22.0'N	083°45.0'E	–	03:22:36.7			16	137	202	226	0.281	0.171
Chelyabinsk	55°10.0'N	061°24.0'E	–	03:12:05.4			4	116	200	232	0.391	0.276
Gorki	56°15.0'N	044°05.0'E	–	03:54 Rise			0	110	–	–	0.092	0.033
Irkutsk	52°16.0'N	104°20.0'E	503	03:44:39.7			25	163	202	213	0.136	0.059
Izevsk	56°51.0'N	053°14.0'E	–	03:19 Rise			0	110	–	–	0.384	0.269
Kazan	55°49.0'N	049°08.0'E	–	03:32 Rise			0	110	–	–	0.328	0.214
Krasnodar	45°02.0'N	039°00.0'E	–	03:51 Rise			0	106	–	–	0.079	0.026
Krasnojarsk	56°01.0'N	092°50.0'E	163	03:33:14.0			18	149	201	219	0.163	0.077
Kujbyshev	53°12.0'N	050°09.0'E	–	03:22 Rise			0	109	–	–	0.440	0.326
Novokuznetsk	53°45.0'N	087°06.0'E	–	03:25:52.3			17	141	202	224	0.247	0.142
Novosibirsk	55°02.0'N	082°55.0'E	–	03:23:31.1			15	137	201	225	0.254	0.148
Omsk	55°00.0'N	073°24.0'E	92	03:17:04.6			10	127	201	229	0.321	0.208
Perm	58°00.0'N	056°15.0'E	–	03:14:18.5			0	112	199	229	0.354	0.239
Rostov-na-Donu	47°14.0'N	039°42.0'E	–	03:52 Rise			0	106	–	–	0.080	0.027
Saratov	51°34.0'N	046°02.0'E	65	03:33 Rise			0	108	–	–	0.355	0.241
Sverdlovsk	48°05.0'N	039°40.0'E	293	03:50 Rise			0	106	–	–	0.117	0.047
Toljatti	53°31.0'N	049°26.0'E	–	03:26 Rise			0	109	–	–	0.410	0.295
Ufa	54°44.0'N	055°56.0'E	187	03:10:12.5			1	111	199	233	0.431	0.317
Vladivostok	43°06.0'N	131°47.0'E	31	04:37:50.8			31	209	200	179	0.075	0.024
Volgograd	48°44.0'N	042°25.0'E	–	03:44 Rise			0	107	–	–	0.213	0.114
SAUDI ARABIA												
Jiddah	21°30.0'N	039°12.0'E	7	03:23 Rise			0	102	–	–	0.239	0.136
Mecca	21°27.0'N	039°49.0'E	–	03:21 Rise			0	102	–	–	0.268	0.160
Medina	24°28.0'N	039°36.0'E	–	03:24 Rise			0	102	–	–	0.278	0.168
Riyadh	24°38.0'N	046°43.0'E	636	02:52 Rise			0	102	–	–	0.657	0.573
SYRIA												
Aleppo	36°12.0'N	037°10.0'E	420	03:43 Rise			0	103	–	–	0.141	0.062
Hims	34°44.0'N	036°43.0'E	–	03:47 Rise			0	104	–	–	0.062	0.019
TURKEY												
Diyarbakir	37°55.0'N	040°14.0'E	–	03:36 Rise			0	104	–	–	0.272	0.163
Gaziantep	37°05.0'N	037°22.0'E	–	03:47 Rise			0	104	–	–	0.085	0.029
Malatya	38°21.0'N	038°19.0'E	–	03:45 Rise			0	104	–	–	0.136	0.059
Maras	37°36.0'N	036°55.0'E	–	03:49 Rise			0	104	–	–	0.049	0.013
Sivas	39°45.0'N	037°02.0'E	–	03:52 Rise			0	105	–	–	0.031	0.006
Urfa	37°08.0'N	038°46.0'E	–	03:41 Rise			0	104	–	–	0.180	0.089
UNITED ARAB EMIRATES												
Abu Dhabi	24°28.0'N	054°22.0'E	–	02:50:00.6			5	105	21	85	0.709	0.636
Dubayy	25°18.0'N	055°18.0'E	–	02:50:20.8			6	106	21	84	0.742	0.676

Table 8b
LOCAL CIRCUMSTANCES DURING THE TOTAL SOLAR ECLIPSE OF 1995 OCTOBER 24 FOR AFRICA, MIDDLE EAST AND RUSSIA

Location Name	First Contact U.T. h m s	Alt °	P °	V °	Second Contact U.T. h m s	Alt °	P °	V °	Third Contact U.T. h m s	Alt °	P °	V °	Fourth Contact U.T. h m s	Alt °	P °	V °
ARMENIA																
Yerevan	−												03:55:29.6	6	117	164
AZERBAIJAN																
Baku	−												03:57:22.9	9	119	165
DJIBOUTI																
Djibouti	−												03:27:56.3	6	60	137
GEORGIA																
Tbilisi	−												03:56:14.5	5	119	165
IRAN																
Ahvaz	−												03:52:30.2	11	104	159
Birjand	−				02:53:21.8	7	157	211	02:53:39.2	7	246	300	04:00:28.0	20	112	161
Esfahan	−												03:54:52.4	13	108	160
Mashhad	−												04:01:40.9	19	118	163
Qom	−												03:55:26.9	12	111	161
Shiraz	−												03:53:51.8	15	103	158
Tabriz	−												03:55:03.6	7	114	163
Tehran	−												03:56:10.3	12	113	162
IRAQ																
Baghdad	−												03:51:48.7	7	105	159
Basra	−												03:51:31.5	11	102	158
Mosul	−												03:53:08.0	5	110	161
KUWAIT																
Kuwait	−												03:50:52.9	11	100	157
QATAR																
Doha	−												03:50:23.3	15	95	155
RUSSIA																
Barnaul	02:34:52.5	11	245	274									04:13:29.4	21	158	176
Chelyabinsk	−												04:03:43.6	10	148	176
Gorki	−												04:01:00.5	1	142	174
Irkutsk	03:06:39.6	23	232	248									04:25:15.7	26	171	175
Izevsk	−												04:02:00.7	5	147	176
Kazan	−												04:01:20.6	3	143	175
Krasnodar	−												03:56:45.5	1	123	166
Krasnojarsk	02:54:23.0	15	234	256									04:14:20.3	20	168	179
Kujbyshev	−												04:01:02.2	5	140	173
Novokuznetsk	02:39:55.7	13	242	270									04:14:46.3	21	160	176
Novosibirsk	02:38:08.7	10	242	270									04:11:38.6	19	159	177
Omsk	02:29:16.2	4	247	279									04:07:46.6	15	154	177
Perm	−												04:02:30.4	6	150	177
Rostov-na-Donu	−												03:57:46.5	1	126	168
Saratov	−												04:00:00.4	3	135	171
Sverdlovsk	−												03:58:06.8	1	128	168
Toljatti	−												04:00:58.5	4	140	173
Ufa	−												04:02:22.2	7	144	175
Vladivostok	04:06:39.0	33	222	207									05:11:11.5	27	176	149
Volgograd	−												03:58:38.1	2	130	169
SAUDI ARABIA																
Jiddah	−												03:40:11.6	3	81	149
Mecca	−												03:40:21.6	4	81	149
Medina	−												03:43:20.2	4	87	152
Riyadh	−												03:46:39.8	11	91	153
SYRIA																
Aleppo	−												03:51:47.6	1	107	160
Hims	−												03:50:47.6	1	105	159
TURKEY																
Diyarbakir	−												03:53:19.8	3	111	162
Gaziantep	−												03:52:22.4	1	109	161
Malatya	−												03:53:15.1	1	111	162
Maras	−												03:52:37.8	1	110	161
Sivas	−												03:53:53.4	0	113	163
Urfa	−												03:52:36.7	2	110	161
UNITED ARAB EMIRATES																
Abu Dhabi	−												03:52:08.4	18	95	154
Dubayy	−												03:53:31.5	19	97	155

Table 9a
CIRCUMSTANCES AT MAXIMUM ECLIPSE ON 1995 OCTOBER 24
FOR AFGHANISTAN AND PAKISTAN

Location Name	Latitude	Longitude	Elev. m	U.T. of Maximum Eclipse h m s	Umbral Duration	Path Width km	Sun Alt °	Sun Azm °	P °	V °	Eclipse Mag.	Eclipse Obs.
AFGHANISTAN												
Farah	32°22.0'N	062°07.0'E	–	02:54:24.5			10	111	202	256	0.998	0.999
Herat	34°20.0'N	062°12.0'E	–	02:55:13.3			9	111	202	254	0.940	0.926
Jalalabad	34°26.0'N	070°28.0'E	–	02:59:43.5			16	117	203	252	0.866	0.833
Kabol	34°52.0'N	069°20.0'E	1954	02:59:07.0			15	116	203	252	0.864	0.830
Konduz	37°45.0'N	068°51.0'E	–	03:00:10.7			14	117	202	249	0.787	0.733
Lashkar Gah	31°35.0'N	064°21.0'E	–	02:55:10.6	00m31.0s	29	12	112	202	256	1.008	1.000
Mazar-e Sharif	36°42.0'N	067°06.0'E	–	02:58:39.4			13	115	202	250	0.832	0.789
Qandahar	31°53.0'N	065°50.0'E	1136	02:56:01.4			13	113	203	256	0.982	0.979
BAHRAIN												
Manama	26°13.0'N	050°35.0'E	–	02:49:36.0			1	103	20	84	0.736	0.669
KAZAKHSTAN												
Alma-Ata	43°15.0'N	076°57.0'E	834	03:08:58.9			18	126	203	240	0.569	0.470
Karaganda	49°50.0'N	073°10.0'E	–	03:11:39.0			12	125	201	234	0.439	0.326
KAZAKHSTA												
Petropavlovsk	54°54.0'N	069°06.0'E	–	03:14:44.8			8	123	200	230	0.351	0.237
KYRGYZSTAN												
Bishkek	42°54.0'N	074°36.0'E	–	03:07:00.1			16	124	203	241	0.599	0.504
PAKISTAN												
Ahmadpur East	29°09.0'N	071°16.0'E	–	02:58:57.6			19	116	24	77	0.999	0.999
Bahawalpur	29°24.0'N	071°41.0'E	–	02:59:18.9			20	116	204	257	1.000	1.000
Chaman	30°55.0'N	066°27.0'E	–	02:56:06.7	00m34.5s	31	14	113	23	77	1.009	1.000
Chiniot	31°43.0'N	072°59.0'E	–	03:00:45.2			20	118	204	254	0.920	0.901
Dera Ghazi Khan	30°03.0'N	070°38.0'E	–	02:58:38.0			18	116	204	256	0.991	0.991
Faisalabad	31°25.0'N	073°05.0'E	–	03:00:46.0			20	118	204	254	0.927	0.911
Gujranwala	32°09.0'N	074°11.0'E	–	03:01:49.3			21	119	204	253	0.895	0.870
Gujrat	32°34.0'N	074°05.0'E	–	03:01:50.3			20	119	204	253	0.885	0.856
Hyderabad	25°29.0'N	068°22.0'E	–	02:56:33.8			18	113	24	82	0.861	0.827
Islamabad	33°42.0'N	073°10.0'E	–	03:01:25.5			19	119	204	252	0.862	0.827
Jhang Sadar	31°16.0'N	072°19.0'E	–	03:00:08.1			19	117	204	255	0.939	0.926
Jhelum	32°56.0'N	073°44.0'E	–	03:01:39.0			20	119	204	252	0.878	0.848
Karachi	24°52.0'N	067°03.0'E	4	02:55:39.7			17	112	23	83	0.830	0.787
Kasur	31°07.0'N	074°27.0'E	–	03:01:50.0			21	119	204	254	0.922	0.904
Lahore	31°35.0'N	074°18.0'E	–	03:01:47.8			21	119	204	254	0.910	0.889
Larkana	27°33.0'N	068°13.0'E	–	02:56:35.2			17	113	23	80	0.921	0.903
Mardan	34°12.0'N	072°02.0'E	–	03:00:44.2			18	118	203	252	0.858	0.823
Mirpur Khas	25°32.0'N	069°00.0'E	–	02:57:00.9			19	113	24	82	0.869	0.837
Multan	30°11.0'N	071°29.0'E	131	02:59:17.1			19	116	204	256	0.979	0.976
Nawabshah	26°15.0'N	068°25.0'E	–	02:56:37.4			18	113	24	81	0.885	0.856
Okara	30°49.0'N	073°27.0'E	–	03:00:56.8			21	118	204	255	0.941	0.928
Peshawar	34°01.0'N	071°33.0'E	–	03:00:19.6			18	118	203	252	0.868	0.835
Quetta	30°12.0'N	067°00.0'E	–	02:56:15.5			15	113	23	77	0.988	0.987
Rahimyar Khan	28°25.0'N	070°18.0'E	–	02:58:08.7			19	115	24	78	0.968	0.962
Rawalpindi	33°36.0'N	073°04.0'E	550	03:01:18.4			19	119	204	252	0.865	0.832
Sahiwal	30°40.0'N	073°06.0'E	–	03:00:38.1			20	118	204	255	0.949	0.938
Sargodha	32°05.0'N	072°40.0'E	–	03:00:35.5			19	118	204	254	0.913	0.892
Shekhupura	31°42.0'N	073°59.0'E	–	03:01:33.4			21	119	204	254	0.910	0.889
Sialkot	32°30.0'N	074°31.0'E	–	03:02:10.8			21	120	204	252	0.882	0.853
Sukkur	27°42.0'N	068°52.0'E	–	02:57:02.7			18	114	23	79	0.932	0.917
Wah Cantonment	33°48.0'N	072°42.0'E	–	03:01:06.0			19	119	203	252	0.863	0.829
TAJIKISTAN												
Dusharibe	38°35.0'N	068°48.0'E	–	03:00:35.7			14	117	202	248	0.764	0.705
TURKMENISTAN												
Ashkhabad	37°37.0'N	058°23.0'E	–	02:55:33.6			6	109	201	251	0.873	0.840
UZBEKISTAN												
Tashkent	41°20.0'N	069°10.0'E	–	03:02:28.2			13	118	202	245	0.686	0.609

Table 9b
**LOCAL CIRCUMSTANCES DURING THE TOTAL SOLAR ECLIPSE OF 1995 OCTOBER 24
FOR AFGHANISTAN AND PAKISTAN**

Location Name	First Contact				Second Contact				Third Contact				Fourth Contact			
	U.T. h m s	Alt °	P °	V °	U.T. h m s	Alt °	P °	V °	U.T. h m s	Alt °	P °	V °	U.T. h m s	Alt °	P °	V °
AFGHANISTAN																
Farah	-												04:02:56.1	23	113	160
Herat	-												04:03:25.3	22	116	162
Jalalabad	01:55:12.2	4	284	338									04:11:51.6	29	121	161
Kabol	01:55:09.5	3	283	337									04:10:30.8	28	121	162
Konduz	01:57:27.9	3	279	330									04:09:46.6	25	125	164
Lashkar Gah	01:52:37.5	0	291	349	02:54:55.2	12	101	155	02:55:26.2	12	303	357	04:05:01.5	25	113	160
Mazar-e Sharif	01:56:04.7	1	281	334									04:08:09.9	25	123	163
Qandahar	01:52:50.9	1	290	347									04:06:38.8	26	115	160
BAHRAIN																
Manama	-												03:50:21.9	14	96	155
KAZAKHSTAN																
Alma-Ata	02:08:14.7	8	266	309									04:15:42.1	26	139	167
Karaganda	02:17:33.2	5	256	294									04:09:55.4	19	146	172
KAZAKHSTA																
Petropavlovsk	02:26:12.0	2	249	282									04:06:13.1	13	151	176
KYRGYZSTAN																
Bishkek	02:06:14.3	6	268	312									04:13:47.0	25	137	167
PAKISTAN																
Ahmadpur East	01:52:57.9	6	293	351									04:13:20.3	33	114	158
Bahawalpur	01:53:08.2	6	292	350									04:13:54.4	33	115	158
Chaman	01:52:34.9	2	291	350	02:55:49.5	14	117	171	02:56:24.0	14	288	341	04:07:15.3	28	113	159
Chiniot	01:54:24.5	7	287	343									04:15:23.3	33	119	159
Dera Ghazi Kh..	01:53:03.1	5	291	349									04:12:25.9	32	115	158
Faisalabad	01:54:18.0	7	288	344									04:15:34.7	33	119	159
Gujranwala	01:55:04.9	8	286	341									04:16:54.6	33	121	159
Gujrat	01:55:16.6	8	285	340									04:16:40.4	33	121	159
Hyderabad	01:52:16.4	4	300	3									04:08:58.7	32	106	156
Islamabad	01:55:37.4	7	284	338									04:15:13.2	31	122	160
Jhang Sadar	01:53:58.6	6	288	345									04:14:34.2	32	118	159
Jhelum	01:55:21.3	7	285	339									04:16:07.3	32	121	160
Karachi	01:52:16.4	3	302	6									04:06:57.5	31	104	155
Kasur	01:54:40.3	8	288	344									04:17:31.9	34	119	158
Lahore	01:54:50.0	8	287	343									04:17:12.5	34	120	159
Larkana	01:52:09.6	4	297	358									04:09:04.6	31	109	157
Mardan	01:55:32.9	6	284	337									04:13:43.7	30	122	161
Mirpur Khas	01:52:20.6	5	300	3									04:09:54.7	33	107	156
Multan	01:53:18.7	6	291	348									04:13:34.0	32	116	158
Nawabshah	01:52:11.9	4	299	1									04:09:10.6	32	107	156
Okara	01:54:10.0	7	289	345									04:16:11.5	34	118	158
Peshawar	01:55:15.5	5	284	338									04:13:11.0	30	122	161
Quetta	01:52:25.9	2	292	351									04:07:49.5	28	113	159
Rahimyar Khan	01:52:36.0	5	294	354									04:12:00.1	33	112	157
Rawalpindi	01:55:30.6	7	284	338									04:15:06.4	31	122	160
Sahiwal	01:53:59.0	7	289	346									04:15:43.9	34	118	158
Sargodha	01:54:29.0	7	287	342									04:14:54.2	32	119	159
Shekhupura	01:54:46.1	8	287	342									04:16:44.7	34	120	159
Sialkot	01:55:24.9	8	285	340									04:17:16.2	33	121	159
Sukkur	01:52:15.6	4	296	357									04:09:58.8	32	110	157
Wah Cantonment	01:55:30.9	6	284	338									04:14:37.1	31	122	160
TAJIKISTAN																
Dushanbe	01:58:16.5	2	277	327									04:09:38.1	25	127	164
TURKMENISTAN																
Ashkhabad	-												04:01:06.6	17	119	164
UZBEKISTAN																
Tashkent	02:01:32.7	3	273	320									04:09:33.6	23	131	166

Table 10a
CIRCUMSTANCES AT MAXIMUM ECLIPSE ON 1995 OCTOBER 24
FOR INDIAN ASIA - I

Location Name	Latitude	Longitude	Elev. m	U.T. of Maximum Eclipse h m s	Umbral Duration	Path Width km	Sun Alt °	Sun Azm °	P °	V °	Eclipse Mag.	Eclipse Obs.
BANGLADESH												
Barisal	22°42.0'N	090°22.0'E	—	03:22:52.1			42	131	208	253	0.966	0.962
Chittagong	22°20.0'N	091°50.0'E	—	03:25:35.0			43	133	208	252	0.956	0.948
Comilla	23°27.0'N	091°12.0'E	—	03:23:53.5			42	133	208	251	0.934	0.921
Dhaka	23°43.0'N	090°25.0'E	—	03:22:27.0			41	132	208	252	0.938	0.925
Jessore	23°10.0'N	089°13.0'E	—	03:20:44.9			40	130	208	254	0.969	0.965
Khulna	22°48.0'N	089°33.0'E	—	03:21:27.9			41	130	208	254	0.975	0.972
Mymensingh	24°45.0'N	090°24.0'E	—	03:21:57.8			40	132	207	251	0.909	0.889
Pabna	24°00.0'N	089°15.0'E	—	03:20:25.7			40	131	207	253	0.946	0.936
Rajshahi	24°22.0'N	088°36.0'E	—	03:19:15.2			39	130	207	253	0.944	0.934
Sylhet	24°54.0'N	091°52.0'E	—	03:24:20.0			41	135	208	249	0.885	0.859
BHUTAN												
Thimbu	27°28.0'N	089°39.0'E	—	03:19:51.0			37	133	207	249	0.846	0.808
INDIA												
Agra	27°11.0'N	078°01.0'E	—	03:04:56.7			27	121	205	257	0.997	0.997
Ahmadabad	23°02.0'N	072°37.0'E	59	03:00:06.4			23	115	25	83	0.833	0.792
Ajmer	26°27.0'N	074°38.0'E	—	03:01:38.5			24	118	25	79	0.956	0.947
Akola	20°44.0'N	077°00.0'E	—	03:05:03.5			29	117	26	84	0.817	0.771
Aligarh	27°53.0'N	078°05.0'E	—	03:04:59.4			26	121	205	257	0.976	0.973
Allahabad	25°27.0'N	081°51.0'E	—	03:09:31.7	00m31.2s	48	31	123	206	257	1.014	1.000
Alwar	27°34.0'N	076°36.0'E	—	03:03:29.8	00m33.3s	42	25	120	205	257	1.012	1.000
Ambala	30°21.0'N	076°50.0'E	—	03:03:51.9			24	121	205	254	0.919	0.900
Amravati	20°56.0'N	077°45.0'E	—	03:05:50.5			30	118	26	84	0.832	0.790
Amritsar	31°35.0'N	074°53.0'E	—	03:02:17.5			22	120	204	253	0.904	0.882
Asansol	23°41.0'N	086°59.0'E	—	03:17:03.3			38	128	207	255	0.985	0.985
Aurangabad	19°53.0'N	075°20.0'E	—	03:03:31.4			28	115	25	86	0.771	0.714
Bangalore	12°58.0'N	077°36.0'E	964	03:09:42.7			34	114	27	92	0.596	0.502
Bareilly	28°21.0'N	079°25.0'E	—	03:06:23.7			27	122	205	255	0.947	0.937
Belgaum	15°52.0'N	074°31.0'E	—	03:04:18.6			29	113	26	90	0.643	0.557
Bhagalpur	25°15.0'N	087°00.0'E	—	03:16:31.5			36	129	207	253	0.941	0.930
Bhavnagar	21°46.0'N	072°09.0'E	—	02:59:54.5			23	114	25	85	0.791	0.738
Bhilai	21°13.0'N	081°26.0'E	—	03:10:14.9			33	121	27	81	0.886	0.860
Bhopal	23°16.0'N	077°24.0'E	—	03:04:48.6			28	118	26	81	0.895	0.871
Bihar	25°11.0'N	085°31.0'E	—	03:14:25.3			35	127	207	254	0.962	0.956
Bikaner	28°01.0'N	073°18.0'E	—	03:00:29.1			22	117	24	78	0.987	0.986
Bokaro Steel City	23°45.0'N	086°07.0'E	—	03:15:44.9			37	127	207	256	0.995	0.996
Bombay	18°58.0'N	072°50.0'E	9	03:01:17.0			25	113	25	88	0.715	0.645
Calcutta	22°32.0'N	088°22.0'E	7	03:19:41.3			40	129	207	255	0.998	0.999
Calicut	11°15.0'N	075°46.0'E	—	03:08:36.1			33	112	26	95	0.522	0.417
Chandigarh	30°44.0'N	076°55.0'E	—	03:03:59.5			24	121	205	253	0.907	0.885
Cochin	09°58.0'N	076°14.0'E	—	03:10:10.2			34	111	27	96	0.490	0.382
Coimbatore	11°00.0'N	076°58.0'E	—	03:10:19.6			34	112	27	95	0.530	0.426
Cuttack	20°30.0'N	085°50.0'E	—	03:16:46.4			38	125	27	79	0.925	0.909
Dehra Dun	30°19.0'N	078°02.0'E	—	03:05:01.8			25	122	205	253	0.907	0.885
Delhi	28°40.0'N	077°13.0'E	—	03:04:07.0			25	120	205	256	0.963	0.956
Dhanbad	23°48.0'N	086°27.0'E	—	03:16:13.1			37	127	207	255	0.989	0.989
Durgapur	23°29.0'N	087°20.0'E	—	03:17:39.7			38	128	207	255	0.986	0.986
Erode	11°21.0'N	077°44.0'E	—	03:11:04.7			35	113	27	94	0.551	0.450
Firozabad	27°09.0'N	078°25.0'E	—	03:05:22.1			27	121	205	257	0.993	0.994
Gaya	24°47.0'N	085°00.0'E	—	03:13:49.1			35	126	207	255	0.980	0.979
Ghaziabad	28°40.0'N	077°26.0'E	—	03:04:20.0			25	121	205	255	0.961	0.953
Gorakhpur	26°45.0'N	083°22.0'E	—	03:11:11.6			32	125	206	254	0.945	0.935
Gulbarga	17°20.0'N	076°50.0'E	—	03:06:14.0			31	115	26	88	0.715	0.644
Guntur	16°18.0'N	080°27.0'E	—	03:11:22.6			35	117	27	87	0.731	0.665
Gwalior	26°13.0'N	078°10.0'E	—	03:05:10.5			27	120	25	78	0.990	0.990
Haora	22°35.0'N	088°20.0'E	—	03:19:36.7			40	129	207	255	0.998	0.999
Hubli-Dharwar	15°21.0'N	075°10.0'E	—	03:05:18.3			30	113	26	91	0.635	0.549
Hyderabad	17°29.0'N	079°28.0'E	571	03:09:23.8			33	117	27	87	0.753	0.691
Indore	22°43.0'N	075°50.0'E	—	03:03:16.1			27	117	25	82	0.861	0.827
Jabalpur	23°10.0'N	079°57.0'E	—	03:07:45.0			31	120	26	80	0.924	0.907

Table 10b
LOCAL CIRCUMSTANCES DURING THE TOTAL SOLAR ECLIPSE OF 1995 OCTOBER 24 FOR INDIAN ASIA - I

Location Name	First Contact U.T. h m s	Alt °	P °	V °	Second Contact U.T. h m s	Alt °	P °	V °	Third Contact U.T. h m s	Alt °	P °	V °	Fourth Contact U.T. h m s	Alt °	P °	V °
BANGLADESH																
Barisal	02:04:09.2	26	293	351									04:52:25.0	54	119	139
Chittagong	02:05:46.2	28	293	350									04:56:14.8	55	120	136
Comilla	02:04:58.6	27	291	348									04:53:34.1	53	121	139
Dhaka	02:04:08.0	26	291	348									04:51:30.3	53	121	140
Jessore	02:02:57.2	25	293	351									04:49:19.1	52	119	141
Khulna	02:03:19.1	26	294	351									04:50:25.6	53	119	140
Mymensingh	02:04:08.9	26	290	345									04:50:23.9	52	122	142
Pabna	02:02:56.7	25	292	348									04:48:36.4	52	120	142
Rajshahi	02:02:19.0	24	292	348									04:46:49.8	51	120	143
Sylhet	02:05:47.0	27	289	343									04:53:27.8	52	124	140
BHUTAN																
Thimbu	02:03:55.1	24	286	339									04:45:58.6	48	125	146
INDIA																
Agra	01:55:03.5	13	293	351									04:24:15.7	40	116	154
Ahmadabad	01:53:32.8	9	303	7									04:15:27.2	38	105	154
Ajmer	01:53:36.3	10	296	356									04:18:43.3	38	112	155
Akola	01:55:45.9	14	305	10									04:23:47.9	44	105	151
Aligarh	01:55:12.8	12	292	349									04:24:07.8	40	117	155
Allahabad	01:57:02.2	17	294	353	03:09:16.5	31	55	106	03:09:47.7	31	355	45	04:32:03.3	45	116	151
Alwar	01:54:28.0	11	293	352	03:03:13.4	25	64	116	03:03:46.8	25	344	36	04:21:44.6	39	115	155
Ambala	01:55:22.7	11	288	344									04:21:15.1	37	120	157
Amravati	01:55:56.9	15	304	8									04:25:19.2	45	106	150
Amritsar	01:55:04.3	9	287	342									04:18:01.3	34	120	158
Asansol	02:00:52.3	23	294	352									04:43:52.3	51	118	145
Aurangabad	01:55:42.1	13	307	13									04:20:25.7	43	102	151
Bangalore	02:03:00.9	19	318	31									04:24:42.3	50	93	146
Bareilly	01:56:01.7	14	290	347									04:26:12.6	41	119	154
Belgaum	01:58:50.0	14	315	25									04:18:07.4	44	95	149
Bhagalpur	02:00:53.7	22	291	348									04:42:37.6	49	120	146
Bhavnagar	01:53:56.9	9	305	11									04:14:30.9	38	103	153
Bhilai	01:57:33.4	18	301	4									04:33:06.3	48	110	148
Bhopal	01:54:58.6	13	300	2									04:24:12.8	43	109	152
Bihar	01:59:38.6	20	292	350									04:39:35.5	48	119	148
Bikaner	01:53:19.0	8	294	353									04:16:24.1	36	114	156
Bokaro Steel ...	02:00:08.6	22	294	353									04:41:55.6	50	117	146
Bombay	01:55:44.7	11	310	18									04:15:20.1	41	99	151
Calcutta	02:02:14.0	25	295	353									04:47:56.3	53	117	142
Calicut	02:05:08.6	18	323	37									04:19:25.1	48	88	146
Chandigarh	01:55:35.2	11	287	343									04:21:14.6	37	120	157
Cochin	02:07:22.0	19	325	40									04:19:58.7	49	86	145
Coimbatore	02:05:46.3	19	323	37									04:22:25.5	50	88	146
Cuttack	02:00:42.3	23	300	2									04:43:31.7	53	112	143
Dehra Dun	01:55:58.1	12	287	343									04:23:06.8	38	121	156
Delhi	01:54:58.8	11	291	348									04:22:25.4	39	118	156
Dhanbad	02:00:24.8	22	294	352									04:42:36.8	50	118	145
Durgapur	02:01:11.4	23	294	352									04:44:48.4	51	118	144
Erode	02:05:24.0	20	322	35									04:24:35.4	51	90	145
Firozabad	01:55:15.0	13	293	351									04:24:58.1	41	116	154
Gaya	01:59:13.5	20	293	351									04:38:48.5	48	118	148
Ghaziabad	01:55:05.0	12	291	348									04:22:46.5	39	118	156
Gorakhpur	01:58:11.4	18	291	348									04:34:13.8	45	119	151
Gulbarga	01:58:00.9	15	311	19									04:23:31.1	46	99	149
Guntur	02:00:17.0	19	311	19									04:32:03.6	51	101	146
Gwalior	01:55:02.2	13	294	354									04:24:50.9	41	115	154
Haora	02:02:11.7	24	295	353									04:47:49.1	53	117	142
Hubli-Dharwar	01:59:31.5	15	315	26									04:19:26.9	45	95	148
Hyderabad	01:58:49.8	18	309	16									04:29:33.1	49	102	147
Indore	01:54:32.9	12	302	5									04:21:18.4	42	107	152
Jabalpur	01:56:10.2	16	299	0									04:29:16.4	45	111	150

Table 11a
CIRCUMSTANCES AT MAXIMUM ECLIPSE ON 1995 OCTOBER 24
FOR INDIAN ASIA - II

Location Name	Latitude	Longitude	Elev. m	U.T. of Maximum Eclipse h m s	Umbral Duration	Path Width km	Sun Alt °	Sun Azm °	P °	V °	Eclipse Mag.	Eclipse Obs.
INDIA												
Jaipur	26°55.0'N	075°49.0'E	—	03:02:44.7			25	119	25	78	0.983	0.981
Jalandhar	31°19.0'N	075°34.0'E	—	03:02:50.4			22	120	204	253	0.905	0.882
Jammu	32°42.0'N	074°52.0'E	—	03:02:31.3			21	120	204	252	0.873	0.842
Jamnagar	22°28.0'N	070°04.0'E	—	02:58:01.1			21	113	24	85	0.789	0.736
Jamshedpur	22°48.0'N	086°11.0'E	—	03:16:13.2			37	126	27	77	0.995	0.996
Jaridih Bazar	23°38.0'N	086°04.0'E	—	03:15:43.1			37	127	207	256	0.999	0.999
Jhansi	25°26.0'N	078°35.0'E	—	03:05:43.1			28	120	25	78	0.972	0.968
Jodhpur	26°17.0'N	073°02.0'E	—	03:00:12.9			22	116	24	80	0.934	0.919
Kakinada	16°56.0'N	082°13.0'E	—	03:13:27.6			37	119	27	86	0.773	0.717
Kanpur	26°28.0'N	080°21.0'E	—	03:07:34.3			29	122	206	256	0.990	0.990
Kharagpur	22°20.0'N	087°20.0'E	—	03:18:09.9			39	127	27	76	0.997	0.998
Kolhapur	16°42.0'N	074°13.0'E	—	03:03:35.3			28	113	26	90	0.664	0.583
Lucknow	26°51.0'N	080°55.0'E	131	03:08:10.9			30	123	206	256	0.972	0.969
Ludhaina	30°54.0'N	075°51.0'E	—	03:03:01.6			23	120	204	254	0.914	0.894
Madras	13°08.0'N	080°15.0'E	17	03:13:13.1			37	116	27	91	0.636	0.550
Madurai	09°56.0'N	078°07.0'E	—	03:12:43.9			36	112	27	96	0.514	0.409
Malegaon	20°33.0'N	074°32.0'E	—	03:02:29.0			26	115	25	85	0.782	0.727
Mangalore	12°52.0'N	074°53.0'E	—	03:06:25.8			31	112	26	94	0.559	0.459
Mathura	27°30.0'N	077°41.0'E	—	03:04:35.2			26	120	205	257	0.991	0.992
Medinipur	22°26.0'N	087°20.0'E	—	03:18:07.1	00m09.4s	54	39	127	27	76	1.015	1.000
Meerut	28°59.0'N	077°42.0'E	—	03:04:36.6			25	121	205	255	0.949	0.938
Mirzapur	25°09.0'N	082°35.0'E	—	03:10:30.7			32	124	206	256	1.014	1.000
Moradabad	28°50.0'N	078°47.0'E	—	03:05:43.1			26	122	205	255	0.941	0.928
Mysore	12°18.0'N	076°39.0'E	—	03:08:58.2			33	113	27	94	0.564	0.465
Nagpur	21°09.0'N	079°06.0'E	—	03:07:21.2			31	119	26	83	0.855	0.819
Nashik	19°59.0'N	073°48.0'E	—	03:01:54.8			26	114	25	86	0.756	0.696
Nellore	14°26.0'N	079°58.0'E	—	03:11:54.2			36	116	27	90	0.670	0.591
New Delhi	28°36.0'N	077°12.0'E	228	03:04:05.7			25	120	205	256	0.965	0.959
Patna	25°36.0'N	085°07.0'E	—	03:13:45.2			34	127	207	254	0.956	0.948
Pondicherry	11°56.0'N	079°53.0'E	—	03:13:36.5			37	115	27	93	0.596	0.503
Pune	18°32.0'N	073°52.0'E	—	03:02:28.0			27	114	25	88	0.714	0.644
Puruliya	23°20.0'N	086°22.0'E	—	03:16:16.5	00m58.6s	53	37	127	207	256	1.015	1.000
Raipur	21°14.0'N	081°38.0'E	—	03:10:30.2			34	121	27	81	0.889	0.864
Rajahmundry	16°59.0'N	081°47.0'E	—	03:12:48.9			36	119	27	86	0.769	0.712
Rajkot	22°18.0'N	070°47.0'E	—	02:58:37.8			22	113	24	85	0.792	0.739
Ranchi	23°21.0'N	085°20.0'E	—	03:14:45.9			36	126	27	77	0.999	0.999
Raurkela	22°13.0'N	084°53.0'E	—	03:14:34.0			36	125	27	78	0.961	0.954
Saharanpur	29°58.0'N	077°33.0'E	—	03:04:31.2			25	121	205	254	0.922	0.905
Salem	11°39.0'N	078°10.0'E	—	03:11:26.1			35	113	27	94	0.565	0.466
Sangli	16°52.0'N	074°34.0'E	—	03:03:53.4			28	114	26	89	0.673	0.594
Solapur	17°41.0'N	075°55.0'E	—	03:05:01.0			29	115	26	88	0.714	0.643
Srinagar	34°05.0'N	074°49.0'E	—	03:02:51.5			20	121	204	251	0.835	0.793
Surat	21°10.0'N	072°50.0'E	—	03:00:39.8			24	114	25	85	0.780	0.725
Tamluk	22°18.0'N	087°55.0'E	—	03:19:05.4	01m09.2s	55	39	128	27	76	1.016	1.000
Tiruchchirappalli	10°49.0'N	078°41.0'E	—	03:12:47.9			36	113	27	94	0.548	0.446
Tirunelveli	08°44.0'N	077°42.0'E	—	03:13:10.5			36	111	27	97	0.474	0.364
Tiruppur	11°06.0'N	077°21.0'E	—	03:10:45.4			34	112	27	95	0.538	0.435
Trivandrum	08°29.0'N	076°55.0'E	—	03:12:18.6			35	111	27	98	0.456	0.345
Tuticorin	08°47.0'N	078°08.0'E	—	03:13:44.4			37	112	27	97	0.481	0.372
Udaipur	24°35.0'N	073°41.0'E	—	03:00:53.1			24	116	25	81	0.891	0.865
Ujjain	23°11.0'N	075°46.0'E	—	03:03:06.3			26	117	25	82	0.874	0.843
Ulhasnagar	19°13.0'N	073°07.0'E	—	03:01:28.7			25	114	25	87	0.726	0.658
Vadodara	22°18.0'N	073°12.0'E	—	03:00:45.9			24	115	25	84	0.818	0.773
Varanasi	25°20.0'N	083°00.0'E	—	03:11:00.2			32	124	206	256	0.990	0.990
Vellore	12°56.0'N	079°08.0'E	—	03:11:48.2			36	115	27	92	0.615	0.525
Vijayawada	16°31.0'N	080°37.0'E	—	03:11:28.5			35	118	27	87	0.740	0.675
Vishakhapatnam	17°42.0'N	083°18.0'E	—	03:14:33.9			38	121	27	84	0.810	0.764
Warangal	18°00.0'N	079°35.0'E	—	03:09:17.5			33	118	27	86	0.769	0.712
NEPAL												
Kathmandu	27°43.0'N	085°19.0'E	1451	03:13:33.3			33	128	206	252	0.894	0.870
Wiratnagar	26°29.0'N	087°17.0'E	—	03:16:34.9			36	130	207	252	0.903	0.881
SRI LANKA												
Colombo	06°56.0'N	079°51.0'E	7	03:18:01.3			40	111	28	98	0.451	0.339

Table 11b
LOCAL CIRCUMSTANCES DURING THE TOTAL SOLAR ECLIPSE OF 1995 OCTOBER 24
FOR INDIAN ASIA - II

Location Name	First Contact U.T. h m s	Alt °	P °	V °	Second Contact U.T. h m s	Alt °	P °	V °	Third Contact U.T. h m s	Alt °	P °	V °	Fourth Contact U.T. h m s	Alt °	P °	V °
INDIA																
Jaipur	01:54:03.9	11	294	354									04:20:36.0	39	114	155
Jalandhar	01:55:14.0	9	287	342									04:19:04.0	35	120	158
Jammu	01:55:41.0	8	285	339									04:17:41.2	34	122	159
Jamnagar	01:53:18.7	7	305	10									04:11:02.8	36	102	154
Jamshedpur	02:00:18.9	22	296	355									04:42:46.3	51	116	145
Jaridih Bazar	02:00:06.8	22	294	353									04:41:54.3	50	117	146
Jhansi	01:55:13.4	14	296	356									04:25:51.8	42	114	153
Jodhpur	01:53:06.6	8	297	358									04:16:08.5	36	111	155
Kakinada	02:00:35.2	21	308	15									04:36:19.2	52	104	145
Kanpur	01:56:12.2	15	293	351									04:28:42.9	43	117	153
Kharagpur	02:01:21.3	24	296	355									04:45:43.7	52	116	143
Kolhapur	01:57:56.0	13	313	23									04:17:40.6	43	96	149
Lucknow	01:56:35.0	15	292	350									04:29:34.4	43	118	152
Ludhaina	01:55:09.8	10	287	343									04:19:36.2	36	120	158
Madras	02:03:41.0	21	316	27									04:31:38.8	53	95	145
Madurai	02:07:48.1	21	324	39									04:25:05.2	52	88	145
Malegaon	01:55:05.9	12	306	12									04:18:51.7	42	103	151
Mangalore	02:02:34.6	16	320	33									04:17:57.1	46	90	147
Mathura	01:54:56.8	12	292	351									04:23:35.1	40	116	155
Medinipur	02:01:20.2	23	296	355	03:18:01.8	39	199	248	03:18:11.2	39	213	262	04:45:38.9	52	116	143
Meerut	01:55:18.3	12	290	347									04:23:05.8	39	118	156
Mirzapur	01:57:30.2	18	294	353									04:33:39.7	46	116	150
Moradabad	01:55:49.2	13	290	346									04:24:56.1	40	119	155
Mysore	02:03:43.6	18	320	33									04:22:06.3	49	91	146
Nagpur	01:56:25.0	16	303	7									04:28:06.1	46	108	150
Nashik	01:55:15.1	11	308	14									04:17:23.1	41	101	151
Nellore	02:02:00.1	20	314	24									04:30:57.1	52	97	145
New Delhi	01:54:57.1	11	291	348									04:22:24.9	39	117	156
Patna	01:59:21.3	20	292	349									04:38:28.0	47	119	148
Pondicherry	02:05:10.6	21	319	31									04:30:31.5	53	93	144
Pune	01:56:16.4	12	310	18									04:17:18.9	42	99	151
Puruliya	02:00:23.3	22	295	354	03:15:47.7	37	80	129	03:16:46.3	37	332	21	04:42:47.1	50	117	145
Raipur	01:57:39.7	18	301	4									04:33:32.3	48	110	148
Rajahmundry	02:00:19.2	20	309	15									04:35:13.9	52	103	145
Rajkot	01:53:28.9	8	305	10									04:12:13.0	37	103	153
Ranchi	01:59:34.0	21	295	355									04:40:30.5	50	116	146
Raurkela	01:59:26.5	21	298	358									04:40:15.4	50	114	146
Saharanpur	01:55:34.5	11	288	344									04:22:29.7	38	120	156
Salem	02:05:03.4	20	321	34									04:25:50.3	51	91	145
Sangli	01:57:50.8	13	313	22									04:18:28.1	44	97	149
Solapur	01:57:27.1	14	311	19									04:21:31.2	45	99	149
Srinagar	01:56:34.8	8	283	335									04:17:12.4	32	124	160
Surat	01:54:22.5	10	306	12									04:15:41.0	39	102	152
Tamluk	02:01:52.3	24	295	355	03:18:30.6	39	142	190	03:19:39.7	40	271	319	04:47:05.8	53	117	142
Tiruchchirapp...	02:06:29.2	21	322	36									04:26:58.6	52	90	145
Tirunelveli	02:09:52.5	21	327	43									04:23:19.9	52	85	144
Tiruppur	02:05:41.9	19	322	36									04:23:28.9	50	89	146
Trivandrum	02:10:13.0	21	328	44									04:20:58.4	51	84	145
Tuticorin	02:09:52.2	22	326	42									04:24:35.4	52	86	144
Udaipur	01:53:26.2	9	299	2									04:17:18.3	38	109	154
Ujjain	01:54:23.3	12	301	4									04:21:07.9	41	108	153
Ulhasnagar	01:55:37.0	11	309	17									04:15:56.3	41	99	151
Vadodara	01:53:56.7	10	304	9									04:16:26.8	39	104	153
Varanasi	01:57:47.1	18	293	352									04:34:23.1	46	117	150
Vellore	02:03:32.5	20	318	29									04:28:40.5	52	94	145
Vijayawada	02:00:09.7	20	310	18									04:32:26.6	51	101	146
Vishakhapatnam	02:00:35.7	22	306	12									04:38:46.4	53	106	144
Warangal	01:58:29.2	18	308	15									04:29:46.0	49	103	148
NEPAL																
Kathmandu	01:59:53.7	19	288	343									04:37:14.7	46	122	150
Wiratnagar	02:01:19.5	22	289	344									04:42:09.2	48	122	147
SRI LANKA																
Colombo	02:13:56.0	25	329	46									04:28:40.6	56	84	143

Table 12a
CIRCUMSTANCES AT MAXIMUM ECLIPSE ON 1995 OCTOBER 24
FOR SOUTHEAST ASIA

Location Name	Latitude	Longitude	Elev. m	U.T. of Maximum Eclipse h m s	Umbral Duration	Path Width km	Sun Alt °	Sun Azm °	P °	V °	Eclipse Mag.	Eclipse Obs.
CAMBODIA												
Angkor Wat	13°26.0'N	103°52.0'E	—	03:59:48.6	01m48.4s	72	62	154	30	56	1.020	1.000
Batdambang	13°06.0'N	103°12.0'E	—	03:58:33.1			62	151	30	58	0.987	0.988
Kampong Cham	12°00.0'N	105°27.0'E	—	04:05:47.1			65	159	30	51	0.993	0.994
Kampong Chhnang	12°15.0'N	104°40.0'E	—	04:03:24.3			64	156	30	54	0.987	0.988
Kampong Saom	10°38.0'N	103°30.0'E	—	04:02:34.3			65	152	30	59	0.925	0.911
Krachen	12°29.0'N	106°01.0'E	—	04:06:37.7	01m60.5s	74	65	161	210	229	1.021	1.000
Phnom Penh	11°33.0'N	104°55.0'E	13	04:04:59.8			65	157	30	53	0.972	0.970
Siemreab	13°22.0'N	103°51.0'E	—	03:59:51.2	01m35.2s	72	62	153	30	56	1.020	1.000
LAOS												
Savannakhet	16°33.0'N	104°45.0'E	—	03:58:14.0			60	157	209	232	0.917	0.901
Vientiane	17°58.0'N	102°36.0'E	183	03:51:23.7			57	151	209	237	0.913	0.895
MYANMAR												
Bago	17°20.0'N	096°29.0'E	—	03:38:19.1			52	138	29	70	0.995	0.996
Dawei	14°05.0'N	098°12.0'E	—	03:45:25.1			57	139	30	70	0.934	0.921
Henzada	17°38.0'N	095°28.0'E	—	03:35:56.1			51	136	29	72	0.987	0.988
Mandalay	22°00.0'N	096°05.0'E	83	03:33:41.5			48	140	208	246	0.903	0.882
Mawlamyine	16°30.0'N	097°38.0'E	49	03:41:35.2			54	139	29	69	0.990	0.991
Monywa	22°05.0'N	095°08.0'E	—	03:31:47.6			47	138	208	248	0.915	0.897
Myingyan	21°28.0'N	095°23.0'E	—	03:32:42.0			47	138	208	248	0.928	0.913
Nyaunglebin	17°57.0'N	096°44.0'E	—	03:38:16.4	01m13.4s	65	52	139	209	249	1.018	1.000
Pathein	16°47.0'N	094°44.0'E	—	03:35:12.6			51	134	29	74	0.953	0.945
Prome	18°49.0'N	095°13.0'E	—	03:34:24.4	01m04.3s	63	49	136	209	251	1.018	1.000
Sittwe	20°09.0'N	092°54.0'E	—	03:28:54.5			46	133	208	253	1.017	1.000
Taunggyi	20°47.0'N	097°02.0'E	—	03:36:29.7			49	141	209	246	0.922	0.906
Yangon	16°47.0'N	096°10.0'E	—	03:38:10.3			52	137	29	71	0.975	0.973
THAILAND												
Bangkok	13°45.0'N	100°31.0'E	17	03:51:10.9			59	144	30	65	0.962	0.957
Chiang Mai	18°47.0'N	098°59.0'E	—	03:42:19.7			53	143	209	244	0.946	0.937
Chon Buri	13°22.0'N	100°59.0'E	—	03:52:45.7			60	145	30	64	0.959	0.953
Hat Yai	07°01.0'N	100°28.0'E	—	03:59:56.2			66	140	30	71	0.777	0.723
Khon Kaen	16°26.0'N	102°50.0'E	—	03:53:38.1			58	151	209	237	0.950	0.942
Lop Buri	14°48.0'N	100°37.0'E	—	03:50:11.4			58	145	30	64	0.991	0.993
Nakhon Pathom	13°49.0'N	100°03.0'E	—	03:49:59.9			59	143	30	66	0.956	0.949
Nakhon Ratchasima	14°58.0'N	102°07.0'E	—	03:53:35.2			59	149	210	240	1.020	1.000
Nakhon Sawan	15°41.0'N	100°07.0'E	—	03:48:01.7	01m44.8s	68	57	144	30	65	1.019	1.000
Pak Chong	14°42.0'N	101°25.0'E	—	03:52:12.6	00m57.8s	70	59	147	30	62	1.020	1.000
Phitsanulok	16°50.0'N	100°15.0'E	—	03:47:05.9			56	145	209	243	0.979	0.978
Phra Nakhon	13°45.0'N	100°31.0'E	—	03:51:10.9			59	144	30	65	0.962	0.957
Sakon Nakhon	17°10.0'N	104°09.0'E	—	03:56:01.9			59	155	209	234	0.910	0.892
Samut Prakan	13°36.0'N	100°36.0'E	—	03:51:33.6			59	144	30	65	0.959	0.953
Samut Sakhon	13°32.0'N	100°17.0'E	—	03:50:53.2			59	144	30	66	0.952	0.945
Saraburi	14°32.0'N	100°55.0'E	—	03:51:12.6			59	146	30	64	0.989	0.990
Songkhla	07°12.0'N	100°36.0'E	—	04:00:00.5			66	141	30	70	0.784	0.732
Takhli	15°15.0'N	100°21.0'E	—	03:49:03.1			58	145	30	64	0.999	1.000
Ubon Ratchathani	15°14.0'N	104°54.0'E	—	04:00:11.3			61	157	209	232	0.950	0.942
Udon Thani	17°26.0'N	102°46.0'E	—	03:52:22.0			57	152	209	237	0.924	0.909
Yala	06°33.0'N	101°18.0'E	—	04:02:42.7			67	142	30	69	0.778	0.724
VIET_NAM												
Bien Hoa	10°57.0'N	106°49.0'E	—	04:10:52.5			67	164	30	46	0.986	0.987
Cam Ranh	11°54.0'N	109°09.0'E	—	04:15:55.2			66	173	209	216	0.974	0.973
Can Tho	10°02.0'N	105°47.0'E	—	04:09:23.7			67	160	30	50	0.945	0.936
Da Nang	16°04.0'N	108°13.0'E	—	04:07:48.5			62	168	209	221	0.878	0.851
Hai Phong	20°52.0'N	106°41.0'E	—	03:58:20.2			56	163	209	225	0.777	0.723
Ha Noi	21°02.0'N	105°51.0'E	7	03:56:06.3			56	160	209	227	0.785	0.733
Ho Chi Minh	10°45.0'N	106°40.0'E	11	04:10:45.1			67	163	30	47	0.978	0.977
Hon Gai	20°57.0'N	107°05.0'E	—	03:59:15.1			56	164	209	224	0.769	0.713
Hue	16°28.0'N	107°42.0'E	—	04:05:56.2			61	166	209	223	0.875	0.847
Long Xuyen	10°23.0'N	105°25.0'E	—	04:07:55.5			66	158	30	52	0.949	0.941
My Tho	10°21.0'N	106°21.0'E	—	04:10:27.9			67	162	30	48	0.963	0.958
Nam Dinh	20°25.0'N	106°10.0'E	—	03:57:30.8			56	161	209	227	0.796	0.747
Nha Trang	12°15.0'N	109°11.0'E	—	04:15:31.3			66	173	209	216	0.964	0.960
Phan Thiet	10°56.0'N	108°06.0'E	—	04:14:23.5	01m47.3s	75	67	169	29	40	1.021	1.000
Qui Nhon	13°46.0'N	109°14.0'E	—	04:13:34.3			64	172	209	217	0.923	0.908
Thai Nguyen	21°36.0'N	105°50.0'E	—	03:55:30.3			55	160	209	227	0.771	0.715
Vinh	18°40.0'N	105°40.0'E	—	03:58:08.0			58	160	209	229	0.848	0.813

Table 12b
LOCAL CIRCUMSTANCES DURING THE TOTAL SOLAR ECLIPSE OF 1995 OCTOBER 24
FOR SOUTHEAST ASIA

Location Name	First Contact U.T. h m s	Alt °	P °	V °	Second Contact U.T. h m s	Alt °	P °	V °	Third Contact U.T. h m s	Alt °	P °	V °	Fourth Contact U.T. h m s	Alt °	P °	V °
CAMBODIA																
Angkor Wat	02:27:35.2	47	298	354	03:58:53.8	62	140	167	04:00:42.2	62	275	301	05:39:50.8	62	116	89
Batdambang	02:26:41.9	46	299	356									05:38:25.1	63	115	90
Kampong Cham	02:31:51.5	50	299	354									05:46:22.1	62	115	81
Kampong Chhna...	02:30:08.5	49	299	355									05:43:47.4	62	115	83
Kampong Saom	02:29:47.6	49	303	2									05:42:31.4	64	112	81
Krachen	02:32:29.3	50	298	352	04:05:43.3	64	93	113	04:07:33.8	65	322	341	05:47:16.1	61	116	81
Phnom Penh	02:31:18.9	50	300	357									05:45:28.1	62	114	80
Siemreab	02:27:36.9	47	298	354	03:59:02.6	62	154	180	04:00:37.8	62	262	288	05:39:53.6	62	116	89
LAOS																
Savannakhet	02:27:03.1	46	292	344									05:37:27.9	59	121	96
Vientiane	02:22:23.4	42	292	344									05:29:28.5	59	122	105
MYANMAR																
Bago	02:13:13.1	36	298	356									05:13:56.3	61	117	119
Dawei	02:17:57.2	40	302	2									05:22:40.9	64	113	107
Henzada	02:11:43.5	35	298	357									05:10:48.9	61	116	122
Mandalay	02:11:04.7	33	290	345									05:06:52.9	56	123	128
Mawlamyine	02:15:17.9	38	298	357									05:18:08.0	62	116	114
Monywa	02:09:47.7	32	291	346									05:04:28.0	56	122	130
Myingyan	02:10:12.8	32	292	347									05:05:52.4	57	122	129
Nyaunglebin	02:13:13.2	36	297	354	03:37:40.7	52	75	116	03:38:54.1	52	340	20	05:13:51.0	60	118	119
Pathein	02:11:20.2	34	300	0									05:09:45.9	61	114	122
Prome	02:10:48.7	34	296	354	03:33:53.3	49	70	112	03:34:57.6	50	345	27	05:08:44.9	59	118	124
Sittwe	02:07:29.6	30	296	354									05:01:14.3	57	118	132
Taunggyi	02:12:36.4	35	292	346									05:10:53.5	58	122	124
Yangon	02:13:08.2	36	299	358									05:13:43.6	62	115	118
THAILAND																
Bangkok	02:21:40.5	43	300	359									05:29:49.4	64	114	99
Chiang Mai	02:16:05.2	38	293	349									05:18:42.8	60	121	116
Chon Buri	02:22:45.5	44	301	359									05:31:40.1	64	114	97
Hat Yai	02:30:02.8	48	312	17									05:36:51.5	70	104	79
Khon Kaen	02:23:34.3	44	294	348									05:32:30.6	60	120	100
Lop Buri	02:20:57.3	42	299	356									05:28:43.7	63	116	102
Nakhon Pathom	02:20:53.8	42	301	360									05:28:23.2	64	114	101
Nakhon Ratcha...	02:23:17.3	44	297	353									05:32:43.4	62	117	98
Nakhon Sawan	02:19:31.1	41	298	355	03:47:09.0	57	133	168	03:48:53.7	57	283	318	05:26:07.5	62	117	105
Pak Chong	02:22:19.4	43	298	355	03:51:42.4	59	177	210	03:52:40.1	59	239	271	05:31:07.8	63	116	99
Phitsanulok	02:18:60.0	40	296	352									05:24:53.5	61	119	108
Phra Nakhon	02:21:40.5	43	300	359									05:29:49.4	64	114	99
Sakon Nakhon	02:25:35.7	45	292	343									05:34:51.7	59	122	99
Samut Prakan	02:21:56.4	43	301	359									05:30:15.4	64	114	98
Samut Sakhon	02:21:30.6	43	301	0									05:29:25.3	64	114	99
Saraburi	02:21:38.6	43	299	356									05:29:56.4	63	116	100
Songkhla	02:29:56.5	48	311	17									05:37:08.2	69	104	79
Takhli	02:20:11.7	42	298	356									05:27:22.1	62	118	104
Ubon Ratchath...	02:28:07.7	47	294	347									05:39:59.6	60	119	92
Udon Thani	02:22:55.7	43	293	345									05:30:46.0	60	121	103
Yala	02:32:01.7	49	312	17									05:39:56.2	69	104	74
VIET NAM																
Bien Hoa	02:35:36.6	52	299	354									05:51:42.2	61	114	74
Cam Ranh	02:39:32.8	54	296	346									05:56:44.0	58	117	73
Can Tho	02:34:38.4	52	302	359									05:49:58.4	62	112	73
Da Nang	02:34:26.9	50	290	337									05:47:22.1	57	123	88
Hai Phong	02:29:41.7	45	284	328									05:34:37.6	55	129	105
Ha Noi	02:27:55.4	44	284	330									05:32:18.1	55	128	107
Ho Chi Minh	02:35:31.8	52	300	355									05:51:33.3	61	114	74
Hon Gai	02:30:33.3	45	283	327									05:35:25.4	54	129	105
Hue	02:33:07.1	49	290	337									05:45:19.8	57	123	90
Long Xuyen	02:33:32.2	51	302	359									05:48:27.6	63	113	75
My Tho	02:35:21.5	52	301	357									05:51:11.5	62	113	73
Nam Dinh	02:28:38.5	45	285	331									05:34:14.1	55	128	105
Nha Trang	02:39:17.9	54	295	345									05:56:16.9	58	117	74
Phan Thiet	02:38:15.2	54	298	351	04:13:28.8	67	148	160	04:15:16.1	67	266	277	05:55:18.5	59	115	71
Qui Nhon	02:38:09.8	53	293	341									05:53:58.0	57	120	79
Thai Nguyen	02:27:52.2	43	283	328									05:31:11.0	55	129	109
Vinh	02:27:58.8	45	288	336									05:36:12.5	57	125	101

Table 13a
CIRCUMSTANCES AT MAXIMUM ECLIPSE ON 1995 OCTOBER 24
FOR THE ORIENT

Location Name	Latitude	Longitude	Elev. m	U.T. of Maximum Eclipse h m s	Umbral Duration	Path Width km	Sun Alt °	Sun Azm °	P °	V °	Eclipse Mag.	Eclipse Obs.
CHINA												
Anshan	41°08.0'N	122°59.0'E	—	04:20:50.1			36	195	202	191	0.161	0.076
Baotou	40°40.0'N	109°59.0'E	—	03:53:25.4			37	170	205	212	0.288	0.178
Beijing	39°55.0'N	116°25.0'E	—	04:07:08.7			38	183	204	202	0.238	0.135
Changchun	43°53.0'N	125°19.0'E	—	04:23:35.5			33	198	201	188	0.101	0.038
Changsha	28°12.0'N	112°58.0'E	53	04:07:14.4			50	178	207	208	0.516	0.411
Chao'an	23°41.0'N	116°38.0'E	—	04:21:41.5			54	190	206	197	0.577	0.480
Chengdu	30°39.0'N	104°04.0'E	—	03:44:52.0			45	158	207	227	0.573	0.476
Chongqing	29°34.0'N	106°35.0'E	281	03:50:55.2			47	163	207	222	0.566	0.467
Dalian	38°53.0'N	121°35.0'E	—	04:19:29.8			39	193	203	193	0.211	0.113
Dongguan	23°03.0'N	113°46.0'E	—	04:14:31.3			55	182	207	205	0.627	0.539
Fushun	41°52.0'N	123°53.0'E	—	04:22:12.8			35	196	202	190	0.142	0.063
Fuzhou	26°06.0'N	119°17.0'E	—	04:26:14.2			51	195	205	191	0.491	0.383
Guangzhou	23°06.0'N	113°16.0'E	19	04:13:06.7			55	181	207	206	0.632	0.545
Guiyang	26°35.0'N	106°43.0'E	—	03:53:17.4			50	163	208	223	0.635	0.549
Hangzhou	30°15.0'N	120°10.0'E	—	04:24:02.1			47	195	205	192	0.392	0.278
Harbin	45°45.0'N	126°41.0'E	156	04:24:47.8			31	199	201	187	0.065	0.020
Jilin	43°51.0'N	126°33.0'E	—	04:26:11.7			32	200	201	187	0.093	0.034
Jinan	36°40.0'N	116°57.0'E	—	04:10:22.2			42	185	204	201	0.294	0.184
Kunming	25°05.0'N	102°40.0'E	2038	03:45:11.0			50	154	208	233	0.728	0.662
Lanzhou	36°03.0'N	103°41.0'E	1675	03:42:09.5			40	158	206	224	0.454	0.342
Nanchang	24°36.0'N	120°59.0'E	—	04:32:48.1			52	201	205	185	0.508	0.402
Nanjing	32°03.0'N	118°47.0'E	—	04:18:38.2			46	190	205	196	0.368	0.254
Pingxiang	27°38.0'N	113°50.0'E	—	04:09:58.8			51	180	207	206	0.519	0.414
Qingdao	36°06.0'N	120°19.0'E	—	04:18:49.6			42	192	204	194	0.273	0.165
Qiqihar	47°19.0'N	123°55.0'E	—	04:18:11.0			30	194	201	192	0.060	0.018
Shanghai	31°14.0'N	121°28.0'E	5	04:26:25.2			46	197	204	189	0.359	0.245
Shenyang	41°48.0'N	123°27.0'E	45	04:21:19.3			35	195	202	191	0.147	0.066
Shijiazhuang	38°03.0'N	114°28.0'E	—	04:03:51.7			40	179	205	205	0.292	0.182
Shuicheng	26°41.0'N	104°50.0'E	—	03:48:53.7			49	159	208	227	0.658	0.577
Suining	30°31.0'N	105°34.0'E	—	03:48:08.0			46	161	207	224	0.557	0.457
Tai'an	36°12.0'N	117°07.0'E	—	04:11:05.4			42	185	204	200	0.302	0.190
Taiyuan	37°55.0'N	112°30.0'E	—	03:59:37.0			40	175	205	209	0.315	0.203
Tangshan	39°38.0'N	118°11.0'E	—	04:11:12.9			39	186	204	199	0.227	0.126
Tianjin	39°08.0'N	117°12.0'E	4	04:09:19.9			39	184	204	200	0.245	0.141
Weifang	36°42.0'N	119°04.0'E	—	04:15:20.2			41	189	204	197	0.273	0.165
Wuhan	30°36.0'N	114°17.0'E	25	04:08:30.2			48	180	206	206	0.447	0.336
Xi'an	34°15.0'N	108°52.0'E	—	03:53:25.1			43	168	206	216	0.432	0.319
Xiaogan	30°55.0'N	113°54.0'E	—	04:07:17.4			48	179	206	206	0.445	0.333
Xintai	35°54.0'N	117°44.0'E	—	04:12:46.5			42	186	204	199	0.301	0.190
Yancheng	33°24.0'N	120°09.0'E	—	04:20:50.7			44	193	204	193	0.327	0.214
Yulin	22°36.0'N	110°07.0'E	—	04:05:17.6			56	172	208	216	0.686	0.610
Zaozhuang	34°53.0'N	117°34.0'E	—	04:13:09.4			43	186	205	199	0.323	0.210
Zhengzhou	34°48.0'N	113°39.0'E	—	04:03:54.1			44	178	205	207	0.366	0.251
Zibo	36°47.0'N	118°01.0'E	—	04:12:47.3			41	187	204	199	0.282	0.172
NORTH KOREA												
P'yongyang	39°01.0'N	125°45.0'E	31	04:29:01.8			37	201	202	185	0.176	0.087
SOUTH KOREA												
Inch'on	37°28.0'N	126°38.0'E	—	04:32:43.0			38	204	202	183	0.197	0.102
Kwangju	35°09.0'N	126°54.0'E	—	04:35:57.4			40	206	202	181	0.238	0.134
Pusan	35°08.0'N	129°05.0'E	2	04:41:19.2			38	210	201	177	0.223	0.123
Seoul	37°33.0'N	126°58.0'E	11	04:33:25.0			38	204	202	182	0.194	0.100
Taegu	35°50.0'N	128°35.0'E	—	04:39:15.4			38	208	202	178	0.214	0.115
MONGOLIA												
Darchan	49°28.0'N	105°56.0'E	—	03:46:12.9			28	165	203	213	0.168	0.081
Ulaanbator	47°55.0'N	106°53.0'E	1406	03:47:25.5			30	166	203	213	0.185	0.093

Table 13b
LOCAL CIRCUMSTANCES DURING THE TOTAL SOLAR ECLIPSE OF 1995 OCTOBER 24
FOR THE ORIENT

Location Name	First Contact U.T. h m s	Alt °	P °	V °	Second Contact U.T. h m s	Alt °	P °	V °	Third Contact U.T. h m s	Alt °	P °	V °	Fourth Contact U.T. h m s	Alt °	P °	V °
CHINA																
Anshan	03:34:13.7	37	235	234									05:10:02.6	33	167	146
Baotou	02:54:55.5	34	248	269									04:56:22.5	37	158	151
Beijing	03:11:34.8	37	244	255									05:06:13.6	36	161	145
Changchun	03:47:10.3	34	227	222									05:02:31.0	30	174	153
Changsha	02:47:26.5	45	267	294									05:32:03.6	46	143	117
Chao'an	02:55:32.8	52	271	294									05:50:33.5	45	138	100
Chengdu	02:27:59.9	36	270	308									05:09:28.5	47	140	134
Chongqing	02:32:33.4	39	270	306									05:16:31.5	48	141	128
Dalian	03:25:55.1	39	240	243									05:15:46.6	35	163	140
Dongguan	02:47:11.9	50	274	304									05:46:04.6	48	136	101
Fushun	03:38:29.1	37	232	230									05:08:28.9	32	169	147
Fuzhou	03:04:42.3	51	264	280									05:49:35.6	42	142	104
Guangzhou	02:45:51.0	50	275	306									05:44:53.2	48	135	102
Guiyang	02:30:58.9	41	275	313									05:23:08.4	50	137	121
Hangzhou	03:10:33.3	47	257	267									05:39:45.1	40	149	116
Harbin	03:56:03.0	32	221	213									04:56:02.5	29	179	159
Jilin	03:51:10.2	34	226	218									05:03:39.8	30	175	153
Jinan	03:08:05.4	40	249	262									05:16:14.3	38	157	136
Kunming	02:22:02.4	38	280	325									05:17:18.2	53	132	123
Lanzhou	02:33:00.9	33	261	294									04:57:53.8	42	148	145
Nanchang	03:09:22.3	53	265	277									05:56:46.8	41	141	98
Nanjing	03:07:57.5	45	255	267									05:32:16.4	40	151	122
Pingxiang	02:49:26.7	46	267	293									05:35:08.8	46	142	114
Qingdao	03:17:44.1	42	247	253									05:22:44.0	37	158	133
Qiqihar	03:50:43.4	31	221	216									04:48:11.5	29	180	165
Shanghai	03:15:40.4	47	254	261									05:39:06.0	38	151	118
Shenyang	03:36:57.3	37	233	232									05:08:15.8	32	169	147
Shijiazhuang	03:02:50.6	38	249	265									05:08:54.9	38	157	142
Shuicheng	02:26:58.9	39	276	317									05:18:57.7	51	136	123
Suining	02:31:04.4	37	269	305									05:12:34.3	47	141	132
Tai'an	03:07:56.0	41	249	262									05:17:48.6	38	156	135
Taiyuan	02:57:06.9	37	251	270									05:06:34.1	39	156	143
Tangshan	03:16:27.1	38	242	251									05:09:13.6	36	162	144
Tianjin	03:12:37.7	38	244	255									05:09:28.1	36	161	143
Weifang	03:14:38.4	41	247	256									05:19:08.9	37	158	135
Wuhan	02:53:04.1	44	261	284									05:28:26.6	44	147	122
Xi'an	02:42:35.5	38	260	288									05:10:06.0	43	149	137
Xiaogan	02:52:17.0	44	261	284									05:26:56.4	44	147	123
Xintai	03:09:23.7	41	249	261									05:19:34.4	38	156	134
Yancheng	03:13:46.3	44	251	260									05:30:34.3	38	154	125
Yulin	02:37:38.7	47	278	316									05:39:04.3	51	133	105
Zaozhuang	03:07:31.9	42	251	264									05:22:14.4	39	155	131
Zhengzhou	02:55:55.3	40	255	275									05:16:24.2	41	152	134
Zibo	03:11:30.4	41	248	258									05:17:25.1	38	158	136
NORTH KOREA																
P'yongyang	03:39:40.6	39	236	232									05:20:30.2	32	165	138
SOUTH KOREA																
Inch'on	03:40:13.9	41	238	232									05:26:58.8	32	163	133
Kwangju	03:37:59.8	43	242	236									05:35:11.7	33	159	125
Pusan	03:45:10.5	42	240	230									05:38:20.3	31	160	124
Seoul	03:41:26.8	41	238	231									05:27:07.7	32	163	133
Taegu	03:44:24.7	42	239	230									05:35:14.2	32	161	126
MONGOLIA																
Darchan	03:03:16.8	25	236	253									04:32:13.4	29	168	170
Ulaanbator	03:01:53.8	27	238	256									04:36:16.9	31	167	167

Table 14a
CIRCUMSTANCES AT MAXIMUM ECLIPSE ON 1995 OCTOBER 24
FOR INDONESIA, JAPAN AND MALAYSIA

Location Name	Latitude	Longitude	Elev. m	U.T. of Maximum Eclipse h m s	Umbral Duration	Path Width km	Sun Alt °	Sun Azm °	P °	V °	Eclipse Mag.	Eclipse Obs.
INDONESIA												
Ambon	03°43.0′S	128°12.0′E	—	05:29:09.3			55	255	21	301	0.807	0.760
Balikpapan	01°17.0′S	116°50.0′E	—	04:57:09.4			72	235	26	330	0.784	0.732
Bandung	06°54.0′S	107°36.0′E	—	04:40:09.8			85	198	29	11	0.499	0.392
Bogor	06°35.0′S	106°47.0′E	—	04:37:31.7			85	181	29	28	0.496	0.389
Cirebon	06°44.0′S	108°34.0′E	—	04:42:28.8			84	212	29	356	0.518	0.413
Denpasar	08°39.0′S	115°13.0′E	—	05:02:29.7			75	257	26	307	0.547	0.446
Jakarta	06°10.0′S	106°48.0′E	9	04:36:56.6			85	180	29	30	0.508	0.403
Jambi	01°36.0′S	103°37.0′E	—	04:21:20.0			78	145	30	66	0.589	0.495
Jember	08°10.0′S	113°42.0′E	—	04:57:58.7			78	253	27	312	0.544	0.443
Kediri	07°49.0′S	112°01.0′E	—	04:53:07.4			80	247	28	319	0.533	0.431
Kudus	06°48.0′S	110°50.0′E	—	04:48:35.7			82	234	28	333	0.547	0.447
Madiun	07°37.0′S	111°31.0′E	—	04:51:32.3			81	244	28	323	0.532	0.430
Magelang	07°28.0′S	110°13.0′E	—	04:47:54.2			83	235	28	332	0.519	0.415
Malang	07°59.0′S	112°37.0′E	—	04:54:55.0			79	250	27	316	0.536	0.434
Manado	01°29.0′N	124°51.0′E	—	05:15:17.1			60	244	23	317	0.941	0.931
Mataram	08°35.0′S	116°07.0′E	—	05:04:42.7			74	258	26	305	0.559	0.460
Medan	03°35.0′N	098°35.0′E	—	04:00:16.6			67	131	30	80	0.651	0.568
Padang	00°57.0′S	100°21.0′E	—	04:11:45.4			73	130	30	81	0.554	0.455
Palembang	02°55.0′S	104°45.0′E	—	04:26:26.2			80	152	30	59	0.570	0.473
Pekalongan	06°53.0′S	109°40.0′E	—	04:45:37.0			83	226	28	342	0.529	0.426
Pekanbaru	00°32.0′N	101°27.0′E	—	04:12:15.5			73	137	30	75	0.614	0.524
Pematangsiantar	02°57.0′N	099°03.0′E	—	04:02:23.5			68	132	31	80	0.641	0.557
Pontianak	00°02.0′N	109°20.0′E	—	04:34:25.5			78	189	29	20	0.721	0.654
Purwokerto	07°25.0′S	109°14.0′E	—	04:45:14.2			84	226	28	342	0.507	0.402
Samarinda	00°30.0′S	117°09.0′E	—	04:56:57.0			71	233	26	331	0.810	0.765
Semarang	06°58.0′S	110°25.0′E	—	04:47:43.5			82	233	28	335	0.537	0.435
Sukabumi	06°55.0′S	106°56.0′E	—	04:38:25.5			85	186	29	24	0.489	0.381
Surabaya	07°15.0′S	112°45.0′E	—	04:54:17.0			79	246	27	320	0.559	0.460
Surakarta	07°35.0′S	110°50.0′E	—	04:49:41.6			82	240	28	327	0.524	0.421
Tahuna	03°37.0′N	125°29.0′E	—	05:14:19.6	01m52.8s	75	58	241	23	320	1.020	1.000
Tanjungkarang-Tel.	05°27.0′S	105°16.0′E	—	04:31:46.5			83	155	30	55	0.506	0.400
Tasikmalaya	07°20.0′S	108°12.0′E	—	04:42:23.3			85	212	29	357	0.495	0.388
Tegal	06°52.0′S	109°08.0′E	—	04:44:10.7			84	220	29	348	0.522	0.418
Ujungpandang	05°07.0′S	119°24.0′E	—	05:08:59.5			69	251	25	311	0.698	0.625
Yogyakarta	07°48.0′S	110°22.0′E	—	04:48:46.1			83	239	28	328	0.512	0.407
JAPAN												
Fukuoka	33°35.0′N	130°24.0′E	—	04:46:30.6			39	213	201	173	0.244	0.140
Hiroshima	34°24.0′N	132°27.0′E	—	04:50:20.9			36	216	200	170	0.218	0.118
Kawasaki	35°32.0′N	139°43.0′E	—	05:04:40.9			30	226	198	161	0.169	0.081
Kitakyushu	33°53.0′N	130°50.0′E	—	04:47:10.2			38	214	201	173	0.236	0.133
Kobe	34°41.0′N	135°10.0′E	—	04:56:13.1			34	220	200	166	0.200	0.104
Kyoto	35°00.0′N	135°45.0′E	—	04:57:03.9			33	221	199	166	0.192	0.098
Nagoya	35°10.0′N	136°55.0′E	—	04:59:22.6			32	223	199	164	0.184	0.093
Osaka	34°40.0′N	135°30.0′E	16	04:56:59.3			34	221	199	166	0.199	0.104
Sapporo	43°03.0′N	141°21.0′E	—	04:56:21.9			25	223	198	167	0.041	0.010
Sendai	38°15.0′N	140°53.0′E	—	05:02:49.0			27	226	198	163	0.118	0.048
Tokyo	35°42.0′N	139°46.0′E	6	05:04:31.7			30	226	198	161	0.166	0.079
Yokohama	35°37.0′N	139°39.0′E	—	05:04:25.2			30	226	198	161	0.168	0.081
MALAYSIA												
Ipoh	04°35.0′N	101°05.0′E	—	04:05:04.1			69	140	30	71	0.720	0.653
Kelang	03°02.0′N	101°27.0′E	—	04:08:22.1			71	140	30	72	0.683	0.607
Kota Baharu	06°08.0′N	102°15.0′E	—	04:05:45.5			68	145	30	66	0.782	0.730
Kota Kinabalu	05°59.0′N	116°04.0′E	—	04:44:29.2			69	212	27	354	0.982	0.982
Kuala Lumpur	03°10.0′N	101°42.0′E	36	04:08:48.6			71	141	30	71	0.691	0.617
Kuala Terengganu	05°20.0′N	103°08.0′E	—	04:09:15.2			70	148	30	63	0.774	0.720
Kuantan	03°48.0′N	103°20.0′E	—	04:12:06.8			72	148	30	63	0.735	0.671
Pinang	05°25.0′N	100°20.0′E	—	04:01:55.5			67	138	30	73	0.731	0.666
Sandakan	05°50.0′N	118°07.0′E	—	04:50:39.6	01m08.6s	78	67	220	26	345	1.021	1.000
Seremban	02°43.0′N	101°56.0′E	—	04:10:06.6			72	141	30	70	0.682	0.606
Taiping	04°51.0′N	100°44.0′E	—	04:03:46.6			68	139	30	72	0.722	0.655

Table 14b
LOCAL CIRCUMSTANCES DURING THE TOTAL SOLAR ECLIPSE OF 1995 OCTOBER 24 FOR INDONESIA, JAPAN AND MALAYSIA

Location Name	First Contact U.T. h m s	Alt °	P °	V °	Second Contact U.T. h m s	Alt °	P °	V °	Third Contact U.T. h m s	Alt °	P °	V °	Fourth Contact U.T. h m s	Alt °	P °	V °
INDONESIA																
Ambon	03:52:43.3	77	304	251									06:52:50.4	35	95	9
Balikpapan	03:17:37.5	76	310	354									06:30:05.6	51	98	21
Bandung	03:13:26.8	70	328	47									06:03:55.1	67	84	4
Bogor	03:11:14.3	68	329	47									06:01:22.1	69	85	6
Cirebon	03:14:28.0	71	327	44									06:07:07.9	66	85	5
Denpasar	03:32:16.0	82	324	33									06:25:24.0	55	84	356
Jakarta	03:09:59.8	68	328	46									06:01:37.0	68	85	8
Jambi	02:52:34.5	59	323	36									05:51:34.7	72	92	34
Jember	03:27:51.8	79	324	37									06:21:37.9	57	85	358
Kediri	03:23:39.7	77	325	40									06:17:05.2	60	85	359
Kudus	03:18:31.8	74	325	39									06:14:05.8	62	86	4
Madiun	03:22:09.6	76	326	41									06:15:44.2	61	85	0
Magelang	03:19:25.5	74	327	43									06:11:55.3	63	85	1
Malang	03:25:15.9	78	325	39									06:18:42.8	59	85	359
Manado	03:34:24.6	77	298	288									06:45:15.2	39	104	26
Mataram	03:33:54.8	83	323	28									06:27:44.5	54	85	356
Medan	02:33:52.1	48	319	30									05:32:41.5	74	97	76
Padang	02:46:33.6	55	325	40									05:40:00.5	76	90	49
Palembang	02:57:40.8	62	324	38									05:55:28.3	71	90	25
Pekalongan	03:16:44.4	72	326	42									06:10:31.1	64	86	4
Pekanbaru	02:44:18.4	55	322	34									05:43:43.5	73	94	50
Pematangsiant...	02:35:46.4	49	320	31									05:34:34.4	74	96	71
Pontianak	02:58:30.5	65	315	19									06:09:28.1	62	98	30
Purwokerto	03:17:37.5	72	328	45									06:08:54.0	65	84	2
Samarinda	03:16:49.1	75	308	350									06:30:30.7	50	100	23
Semarang	03:18:16.8	73	325	41									06:12:46.5	62	86	3
Sukabumi	03:12:29.4	69	329	48									06:01:41.8	68	84	5
Surabaya	03:23:22.7	77	324	35									06:19:27.2	59	86	1
Surakarta	03:20:50.7	75	326	42									06:13:43.6	62	85	0
Tahuna	03:33:16.7	75	294	284	05:13:22.6	58	124	62	05:15:15.4	58	278	214	06:44:39.8	37	108	32
Tanjungkarang...	03:05:36.1	65	328	46									05:56:42.3	71	86	13
Tasikmalaya	03:15:42.8	71	328	47									06:05:38.6	66	84	2
Tegal	03:15:47.7	72	327	43									06:08:51.8	65	85	4
Ujungpandang	03:32:20.7	83	314	343									06:36:34.6	48	92	7
Yogyakarta	03:20:41.6	74	327	44									06:12:11.8	63	84	360
JAPAN																
Fukuoka	03:47:41.9	44	242	229									05:45:29.6	31	158	118
Hiroshima	03:55:12.3	42	239	222									05:45:35.3	29	159	120
Kawasaki	04:17:50.7	36	232	203									05:51:09.0	22	162	119
Kitakyushu	03:49:26.3	43	241	227									05:45:05.7	30	158	119
Kobe	04:03:56.7	40	236	215									05:48:18.4	26	160	119
Kyoto	04:06:02.7	39	235	213									05:47:56.7	26	161	119
Nagoya	04:09:39.2	38	234	210									05:48:53.0	25	161	119
Osaka	04:04:55.9	40	236	214									05:48:48.9	26	160	118
Sapporo	04:34:21.8	27	214	186									05:20:04.8	22	180	146
Sendai	04:24:11.1	32	226	197									05:41:55.9	22	168	127
Tokyo	04:18:09.7	36	231	203									05:50:35.2	22	163	119
Yokohama	04:17:45.5	36	232	203									05:50:45.1	22	162	119
MALAYSIA																
Ipoh	02:35:12.1	51	315	23									05:40:40.0	71	101	68
Kelang	02:38:47.0	52	317	27									05:42:47.2	72	98	61
Kota Baharu	02:34:10.0	51	311	17									05:43:20.3	69	104	69
Kota Kinabalu	03:03:08.7	67	299	338									06:22:35.1	50	110	45
Kuala Lumpur	02:38:52.0	53	317	26									05:43:33.6	71	99	60
Kuala Terengg...	02:36:55.9	52	312	17									05:46:49.1	68	103	63
Kuantan	02:40:04.3	54	314	21									05:48:34.8	69	101	57
Pinang	02:32:36.3	49	314	22									05:37:37.5	71	101	74
Sandakan	03:08:50.6	70	297	329	04:50:03.3	67	172	131	04:51:11.9	67	237	195	06:27:32.1	48	111	43
Seremban	02:40:07.8	53	317	27									05:44:34.5	71	98	57
Taiping	02:34:12.3	50	315	23									05:39:19.8	71	101	70

Table 15a
CIRCUMSTANCES AT MAXIMUM ECLIPSE ON 1995 OCTOBER 24
FOR AUSTRALIA, THE PHILIPPINES AND PACIFIC

Location Name	Latitude	Longitude	Elev. m	U.T. of Maximum Eclipse h m s	Umbral Duration	Path Width km	Sun Alt °	Sun Azm °	P °	V °	Eclipse Mag.	Eclipse Obs.
AUSTRALIA												
Alice Springs	23°42.0′S	133°53.0′E	588	05:48:60.0			46	278	18	266	0.191	0.098
Brisbane	27°28.0′S	153°02.0′E	6	06:06:37.6			24	269	13	258	0.058	0.017
Cairns	16°55.0′S	145°46.0′E	—	06:03:21.6			31	267	14	272	0.412	0.298
Darwin	12°25.0′S	131°00.0′E	—	05:41:19.0			51	267	19	284	0.547	0.446
Gladstone	23°51.0′S	151°16.0′E	—	06:06:50.7			26	268	13	262	0.178	0.088
Ipswich	27°36.0′S	152°46.0′E	—	06:06:24.9			24	269	13	258	0.055	0.015
Rockhampton	23°23.0′S	150°31.0′E	—	06:06:27.7			26	268	13	263	0.194	0.100
Southport	27°58.0′S	153°25.0′E	—	06:06:37.6			24	269	13	258	0.042	0.010
Stirling	21°44.0′S	133°45.0′E	—	05:48:40.9			46	276	18	269	0.255	0.149
Toowoomba	27°33.0′S	151°57.0′E	—	06:05:56.9			25	270	13	258	0.059	0.017
BRUNEI DARUSSALAM												
Bandar Seri Begawan	04°56.0′N	114°55.0′E	3	04:42:43.0			71	210	28	357	0.938	0.927
FIJI												
Suva	18°08.0′S	178°25.0′E	7	06:09 Set			0	257	—	—	0.219	0.119
GUAM												
Agana	13°28.0′N	144°47.0′E	118	05:45:19.7			30	248	195	128	0.673	0.594
HONG KONG												
Victoria	22°17.0′N	114°09.0′E	36	04:16:26.6			56	184	207	203	0.640	0.555
MACAU												
Macau	22°12.0′N	113°32.0′E	—	04:14:50.9			56	182	207	205	0.650	0.567
MICRONESIA												
Kolonia	06°58.0′N	158°13.0′E	—	06:06:27.5			14	256	192	112	0.893	0.866
NEW CALEDONIA												
Noumea	22°18.0′S	166°48.0′E	81	06:13:35.7			10	261	11	260	0.163	0.077
PAPUA NEW GUINEA												
Lae	06°45.0′S	147°00.0′E	—	06:03:02.8			29	260	14	283	0.735	0.669
Port Moresby	09°30.0′S	147°10.0′E	41	06:03:59.7			29	262	14	280	0.647	0.563
PHILIPPINES												
Angeles	15°09.0′N	120°35.0′E	—	04:44:24.0			59	211	205	175	0.742	0.679
Bacolod	10°40.0′N	122°57.0′E	—	04:57:44.8			59	224	204	160	0.837	0.798
Butuan	08°57.0′N	125°33.0′E	—	05:07:31.2			57	232	203	150	0.861	0.829
Cabanatuan	15°29.0′N	120°58.0′E	—	04:45:03.3			59	212	205	174	0.729	0.664
Cagayan de Oro	08°29.0′N	124°39.0′E	—	05:05:38.6			58	231	204	151	0.881	0.855
Cavite	14°29.0′N	120°55.0′E	—	04:46:20.1			59	213	205	173	0.756	0.696
Cebu	10°18.0′N	123°54.0′E	—	05:00:59.8			58	227	204	157	0.838	0.800
Davao	07°04.0′N	125°36.0′E	29	05:10:11.3			57	235	203	146	0.913	0.895
General Santos	06°07.0′N	125°11.0′E	—	05:10:18.2			58	236	203	145	0.943	0.933
Iloilo	10°42.0′N	122°34.0′E	—	04:56:35.4			60	223	205	161	0.839	0.801
Manila	14°35.0′N	121°00.0′E	16	04:46:26.2			59	213	205	173	0.752	0.692
Olongapo	14°50.0′N	120°16.0′E	—	04:43:55.1			60	210	206	175	0.753	0.694
Quezon City	14°38.0′N	121°03.0′E	—	04:46:30.8			59	213	205	173	0.750	0.690
Zamboanga	06°54.0′N	122°04.0′E	—	05:00:31.2			62	229	205	155	0.948	0.940
SAIPAN												
Susupe	15°09.0′N	145°43.0′E	—	05:44:55.0			29	247	195	130	0.628	0.539
SINGAPORE												
Singapore	01°17.0′N	103°51.0′E	11	04:17:24.0			75	149	30	62	0.674	0.596
SOLOMON IS.												
Honiara	09°26.0′S	159°57.0′E	—	06:13:31.6			14	260	11	274	0.603	0.510
TAIWAN												
Kaohsiung	22°38.0′N	120°17.0′E	—	04:33:18.3			54	201	205	185	0.561	0.462
T'aichung	24°09.0′N	120°41.0′E	—	04:32:31.0			52	201	205	186	0.521	0.417
T'aipei	25°03.0′N	121°30.0′E	7	04:33:41.7			51	202	205	184	0.492	0.385
VANUATU												
Port Vila	17°44.0′S	168°19.0′E	—	06:15:06.7			7	260	10	264	0.297	0.186

Table 15b
LOCAL CIRCUMSTANCES DURING THE TOTAL SOLAR ECLIPSE OF 1995 OCTOBER 24 FOR AUSTRALIA, THE PHILIPPINES AND PACIFIC

Location Name	First Contact U.T. h m s	Alt °	P °	V °	Second Contact U.T. h m s	Alt °	P °	V °	Third Contact U.T. h m s	Alt °	P °	V °	Fourth Contact U.T. h m s	Alt °	P °	V °
AUSTRALIA																
Alice Springs	04:58:11.7	57	342	226									06:33:19.5	36	51	301
Brisbane	05:42:36.9	30	353	238									06:28:25.7	19	32	277
Cairns	04:59:09.2	47	321	219									06:59:18.3	18	66	322
Darwin	04:18:13.0	71	318	223									06:52:50.5	33	78	341
Gladstone	05:25:14.0	35	339	228									06:44:18.5	17	47	295
Ipswich	05:43:02.2	30	354	239									06:27:36.5	20	31	276
Rockhampton	05:22:44.1	36	337	227									06:45:42.1	17	48	297
Southport	05:46:18.8	28	356	241									06:25:00.5	20	29	273
Stirling	04:50:33.7	59	337	224									06:39:18.4	34	57	309
Toowoomba	05:41:36.8	31	353	238									06:27:57.9	20	32	277
BRUNEI DARUSSALAM																
Bandar Seri B..	03:01:43.6	67	301	346									06:20:55.9	52	108	42
FIJI																
Suva	05:38:03.6	7	331	224									–			
GUAM																
Agana	04:23:57.0	47	267	212									06:55:57.4	14	122	49
HONG KONG																
Victoria	02:48:05.9	51	275	305									05:48:40.6	48	135	98
MACAU																
Macau	02:46:23.8	51	276	307									05:47:31.8	49	134	99
MICRONESIA																
Kolonia	04:53:18.7	32	276	202									–			
NEW CALEDONIA																
Noumea	05:38:53.6	18	338	228									06:45:57.4	3	43	291
PAPUA NEW GUINEA																
Lae	04:44:42.9	48	301	214									07:09:55.0	13	86	352
Port Moresby	04:48:47.6	47	306	214									07:08:29.6	13	81	344
PHILIPPINES																
Angeles	03:07:57.5	62	280	298									06:18:02.7	43	126	70
Bacolod	03:18:01.8	67	285	295									06:30:57.1	40	119	54
Butuan	03:27:40.5	69	286	282									06:38:27.2	37	117	47
Cabanatuan	03:09:01.3	62	280	296									06:18:09.6	43	127	70
Cagayan de Oro	03:25:14.2	70	287	288									06:37:29.4	38	116	47
Cavite	03:09:17.3	63	281	298									06:20:02.7	43	125	67
Cebu	03:21:18.7	68	285	290									06:33:21.9	39	119	53
Davao	03:29:36.4	71	289	283									06:41:06.2	37	114	42
General Santos	03:29:19.2	72	290	286									06:41:31.1	38	112	39
Iloilo	03:16:48.2	67	286	297									06:30:08.8	41	119	55
Manila	03:09:30.1	63	281	297									06:20:01.3	43	125	68
Olongapo	03:07:06.9	62	281	300									06:18:01.0	43	125	69
Quezon City	03:09:38.1	63	281	297									06:20:01.5	43	125	68
Zamboanga	03:18:54.3	71	292	305									06:34:39.2	42	113	43
SAIPAN																
Susupe	04:25:51.6	45	264	210									06:53:53.9	13	125	53
SINGAPORE																
Singapore	02:46:04.2	57	318	27									05:51:38.4	70	97	45
SOLOMON IS.																
Honiara	05:09:18.2	30	306	211									07:10:00.1	1	77	337
TAIWAN																
Kaohsiung	03:06:26.9	55	269	284									06:00:29.0	42	137	93
T'aichung	03:08:13.3	53	266	279									05:57:23.1	41	140	97
T'aipei	03:11:14.9	53	264	274									05:56:33.5	40	142	99
VANUATU																
Port Vila	05:30:06.4	18	326	221												

Table 16

CLIMATOLOGICAL STATISTICS ALONG THE ECLIPSE PATH OF THE TOTAL SOLAR ECLIPSE OF 1995 OCTOBER 24

Location	Days with scattered cloud and good visibility at eclipse time	Hours of sunshine	Percent of possible sunshine	Mean daily cloudiness (10ths)	Days with rain	Days with fog	Days with dust	Days with thunder-storms
Iran								
Teheran	19.5				0.8			0.1
Hamadan	19.3				1.7			1.2
Birjand	24.7							0.0
Afghanistan								
Herat	-				0.3			0.4
Farah	-				0.2			0.0
Ghazni	-				0.3			
Khandahar	28.8	9.9	89	2.5	0			0.0
Pakistan								
Fort Sandeman	31.0				0.3			0.0
Quetta	24.7			1.1	0.3	0		0.1
Multan	29.2			0.6	0.2	0		0.3
Jacobabad	27.9				0.2			0.0
India								
Bikaner	28.7			0.5	0.6	0		0.4
Jodhpur	27.4	10.3	89		0.6			1.0
Jaipur	25.9	9.6	83	1.4	1.1	<1	0	2.0
Delhi	23.6	9.1	79	1.1	1.0	<1	<1	1.0
Agra	28.6	9.2	80	1.1	1.5	0.1	<1	1.0
Gwalior	24.2				5.9			
Lucknow	21.4			1.5	2.5	<1	<1	0.3
Allahabad	21.0	8.9	77	2.6	3.0			1.0
Patna		8.6	74	2.9	4.0	<1	0	5.0
Asansol	11.4				6.0			4.0
Jamshedpur	11.7				4.0			5.0
Calcutta	7.4	6.3	54	4.9	9.3	0.6	0	5.9
Bangladesh								
Jessore	5.0				6.3			0.1
Myanmar								
Sandoway	5.0				13.2			10.2
Toungoo	2.2				10.4			
Bassein	0.3				10.8			11.0
Mingaladon	0.6				9.1			8.5
Moulmein	3.5				11.6			
Thailand								
Ban Mae Sot	2.3				7.3			8.8
Phitsanulok	2.6				9.0			12.0
Koke Kathiem	1.7				9.5			9.0
Korat	2.2				9.9			6.4
Nakhon Ratchasima					11			
Chaiyaphum	2.6				6.6			
Surin	1.1				8.7			14.0
Aranyaprathet	2.1				10.8			9.0
Cambodia								
Battambang	4.3				11.8			7.9
Siem Riep	0.9				12.2			
Stung Treng	5.3				10.6			4.0
Phnom Penh	1.3	6.5	55	7.8	12.7			5.0
Kompong Cham	2.9				12.2			15.0
Kratie		6.0	50	8.9	12			
Vietnam								
Bien Hoa	0.0				11.2			8.0
Dalat	0.5				12.6			4.8
Ho Chi Minh City	0.1	4.5	38	9.3	12.9			6.5
Phan Thiet	5.2				10.4			5.0
Borneo								
Kinabalu	0.4				15.4			6.4
Sandakan	0.4				12.9			
Philippines								
Jolo Bay	0.6			7.5	11.5			0.0
Zamboanga	1.9			7.5	7.9			2.9
Indonesia								
Mapanget	3.5				8.2			4.4
Pitu	9.3				7.3			7.2
Caroline Is.								
Koror Island	0.2				15.9			5.3
Moen Flight Strip	0.8				17.0			2.7
Marshall Is.								
Kwajalein KTS	0.5				16.8			0.6

Table 17

SOLAR ECLIPSE EXPOSURE GUIDE

ISO				f/Number					
25		1.4	2	2.8	4	5.6	8	11	16	22
50		2	2.8	4	5.6	8	11	16	22	32
100		2.8	4	5.6	8	11	16	22	32	44
200		4	5.6	8	11	16	22	32	44	64
400		5.6	8	11	16	22	32	44	64	88
800		8	11	16	22	32	44	64	88	128
1600		11	16	22	32	44	64	88	128	176

Subject	Q				Shutter Speed				
Solar Eclipse									
Partial[1] - 4.0 ND	11	—	—	—	1/4000	1/2000	1/1000	1/500	1/250	1/125
Partial[1] - 5.0 ND	8	1/4000	1/2000	1/1000	1/500	1/250	1/125	1/60	1/30	1/15
Baily's Beads[2]	12	—	—	—	—	1/4000	1/2000	1/1000	1/500	1/250
Chromosphere	11	—	—	—	1/4000	1/2000	1/1000	1/500	1/250	1/125
Prominences	9	—	1/4000	1/2000	1/1000	1/500	1/250	1/125	1/60	1/30
Corona - 0.1 Rs	7	1/2000	1/1000	1/500	1/250	1/125	1/60	1/30	1/15	1/8
Corona - 0.2 Rs[3]	5	1/500	1/250	1/125	1/60	1/30	1/15	1/8	1/4	1/2
Corona - 0.5 Rs	3	1/125	1/60	1/30	1/15	1/8	1/4	1/2	1 sec	2 sec
Corona - 1.0 Rs	1	1/30	1/15	1/8	1/4	1/2	1 sec	2 sec	4 sec	8 sec
Corona - 2.0 Rs	0	1/15	1/8	1/4	1/2	1 sec	2 sec	4 sec	8 sec	15 sec
Corona - 4.0 Rs	-1	1/8	1/4	1/2	1 sec	2 sec	4 sec	8 sec	15 sec	30 sec
Corona - 8.0 Rs	-3	1/2	1 sec	2 sec	4 sec	8 sec	15 sec	30 sec	1 min	2 min

Exposure Formula: $t = f^2 / (I \times 2^Q)$ where: t = exposure time (sec)
 f = f/number or focal ratio
 I = ISO film speed
 Q = brightness exponent

Abbreviations: ND = Neutral Density Filter.
 Rs = Solar Radii.

Notes: [1] Exposures for partial phases are also good for annular eclipses.
 [2] Baily's Beads are extremely bright and change rapidly.
 [3] This exposure is also recommended for the 'Diamond Ring' effect.

F. Espenak - Nov 1992

TOTAL SOLAR ECLIPSE OF 1995 OCTOBER 24

Maps of the Umbral Path

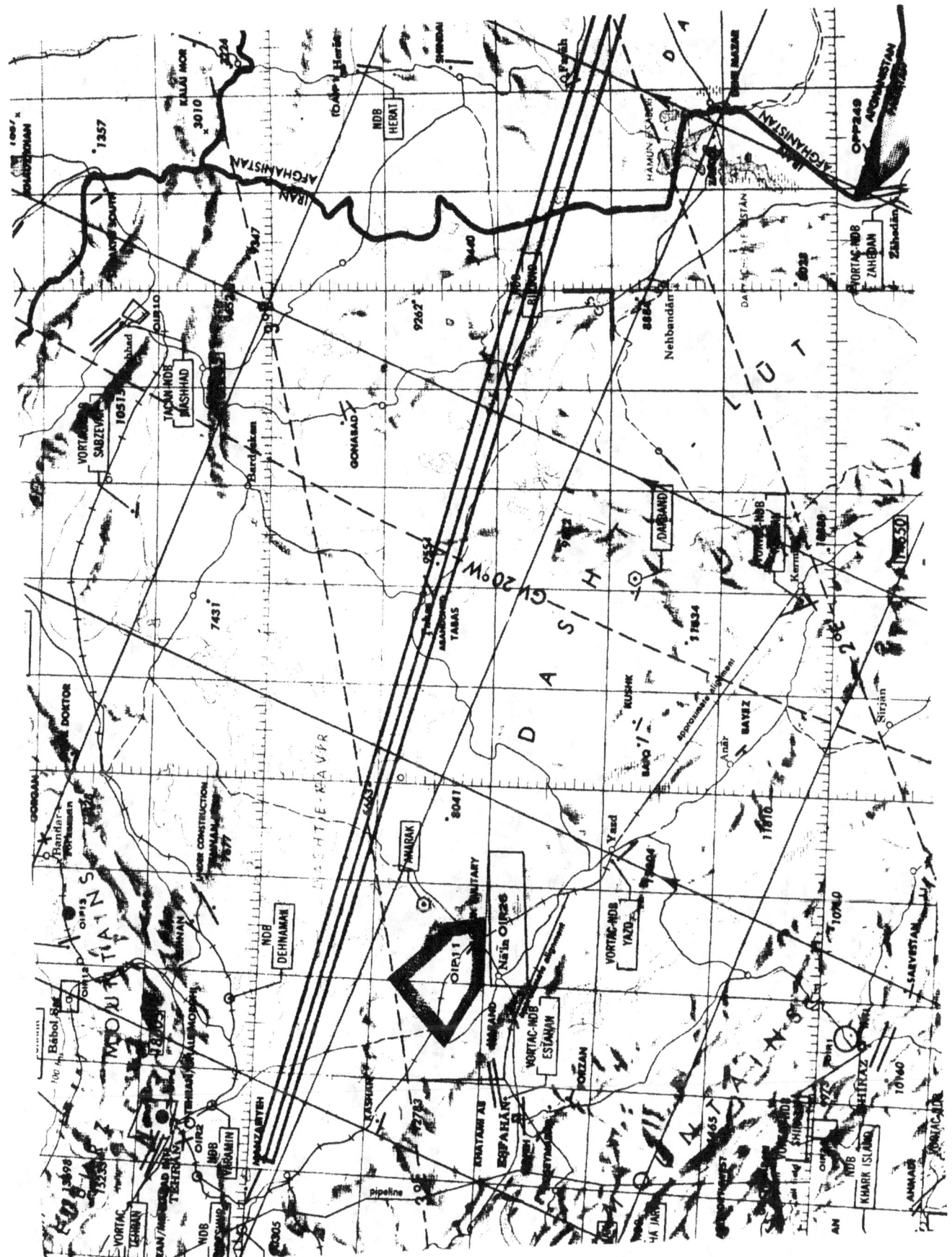

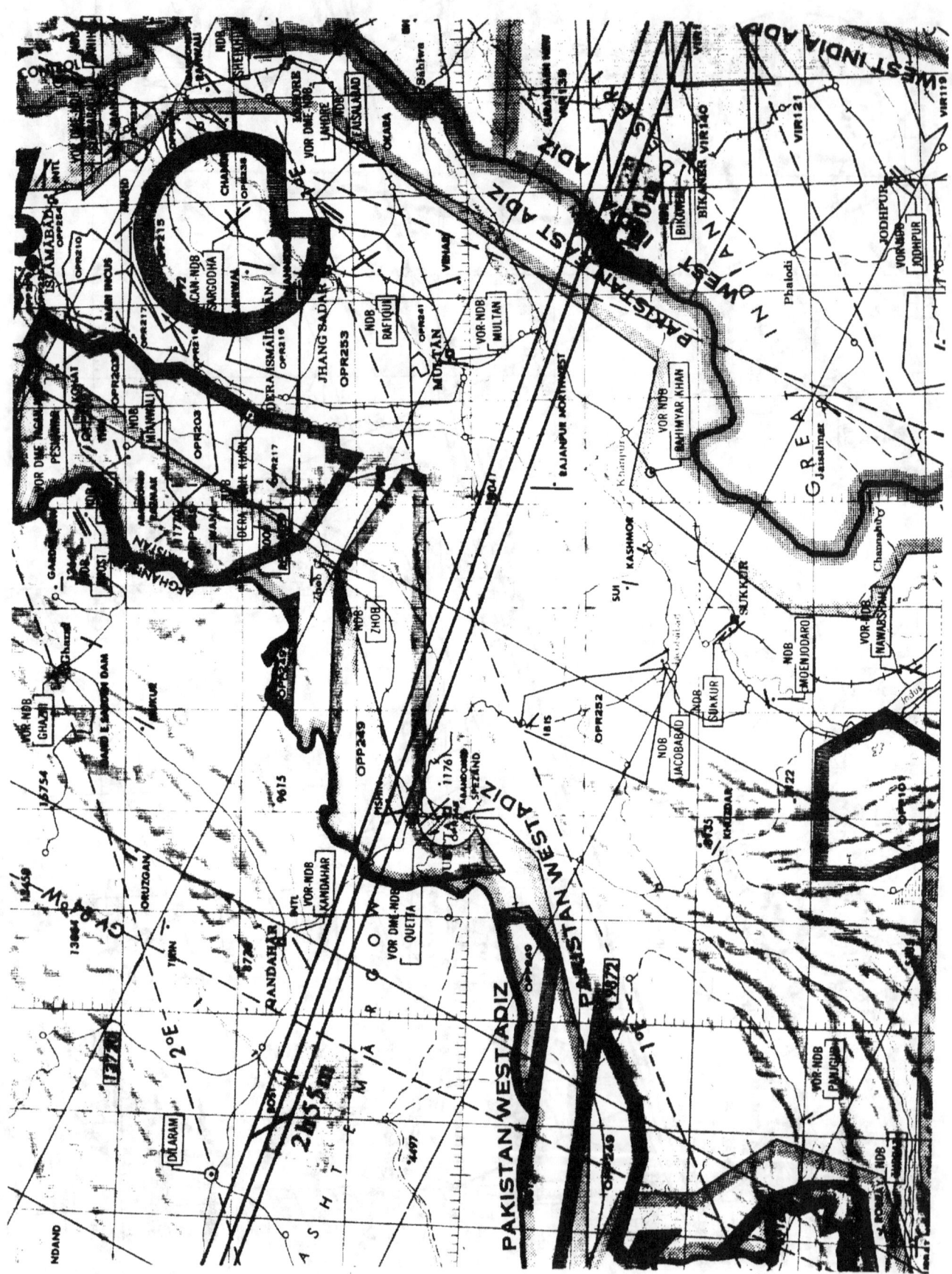

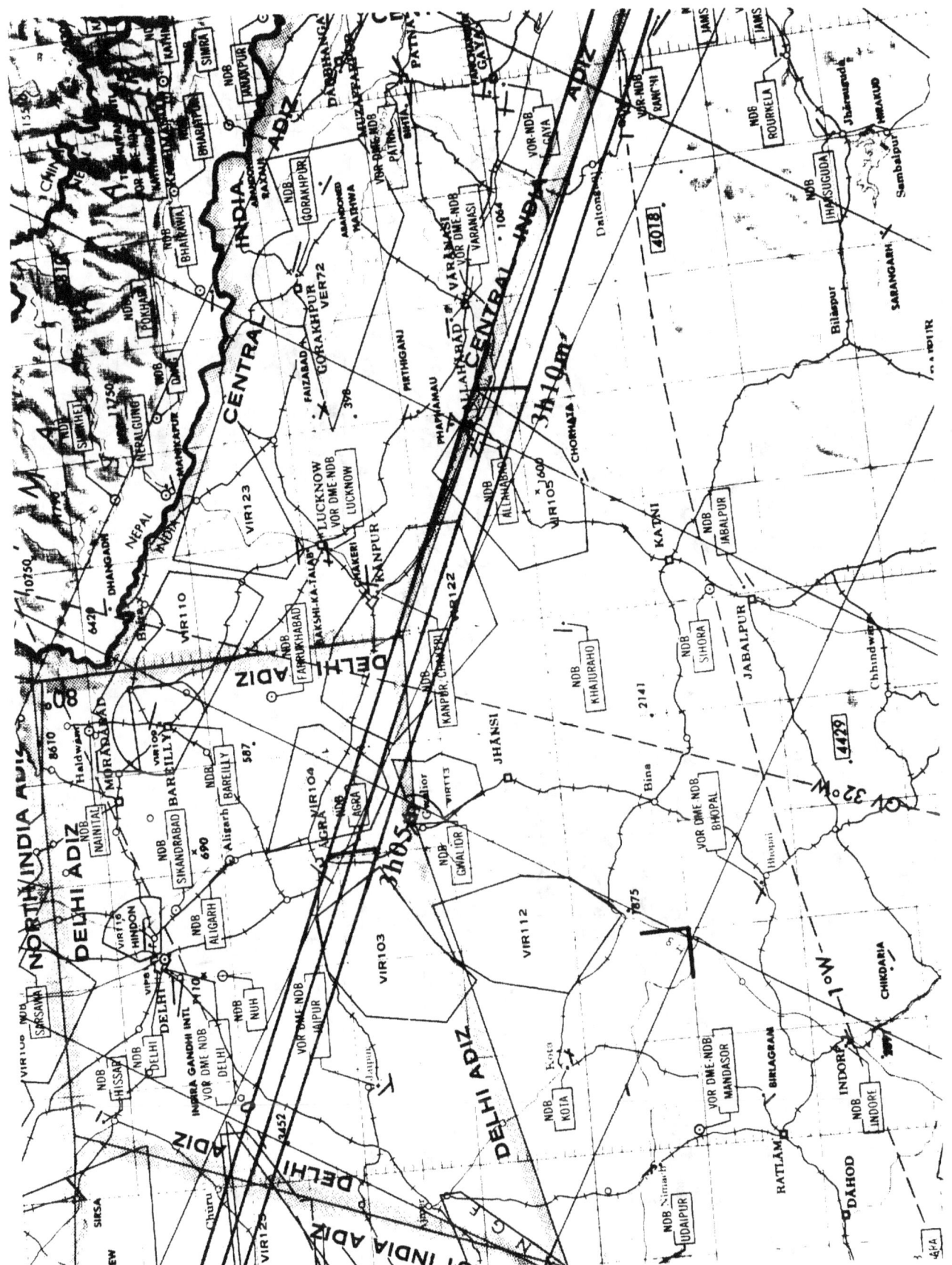

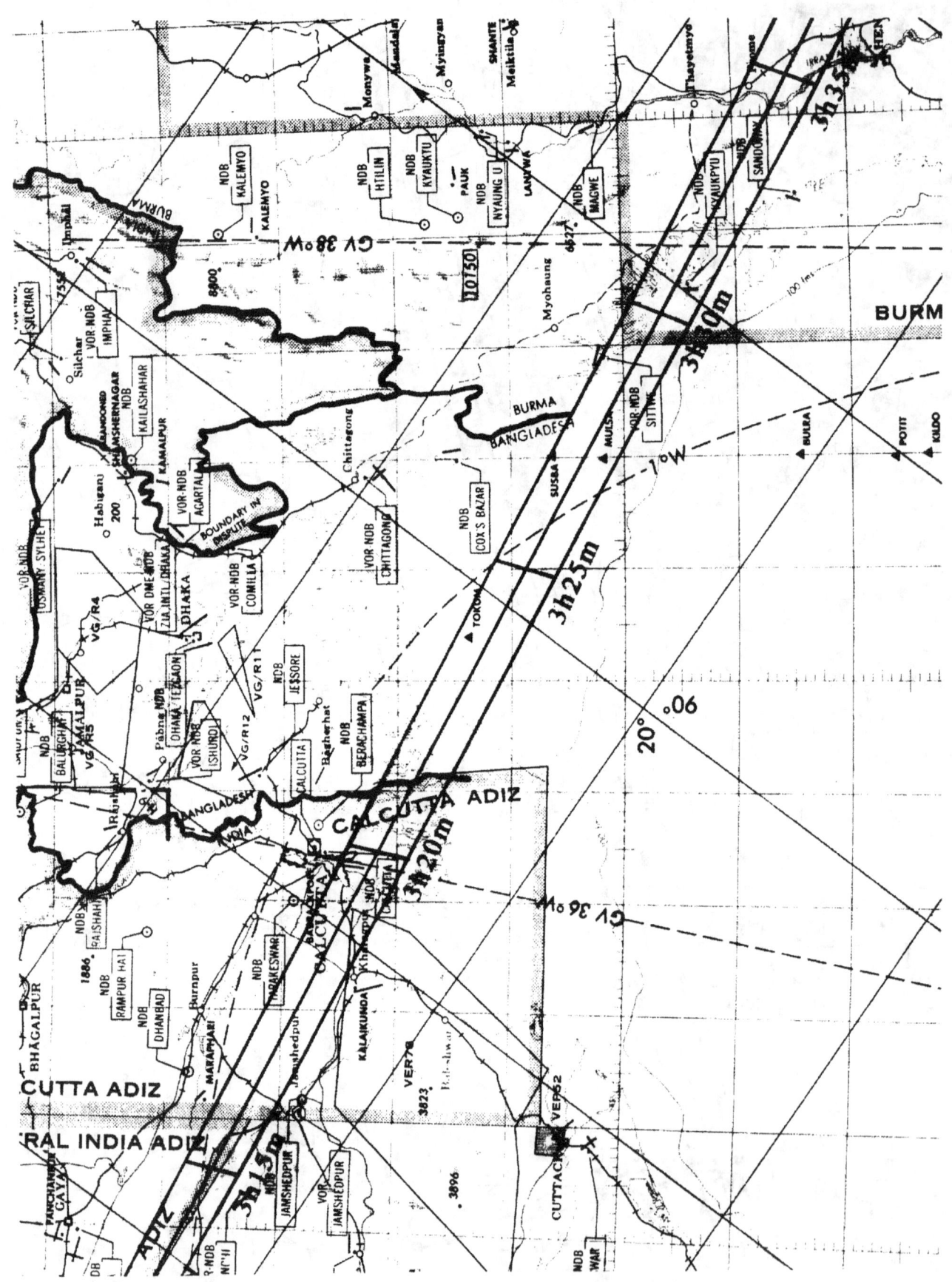

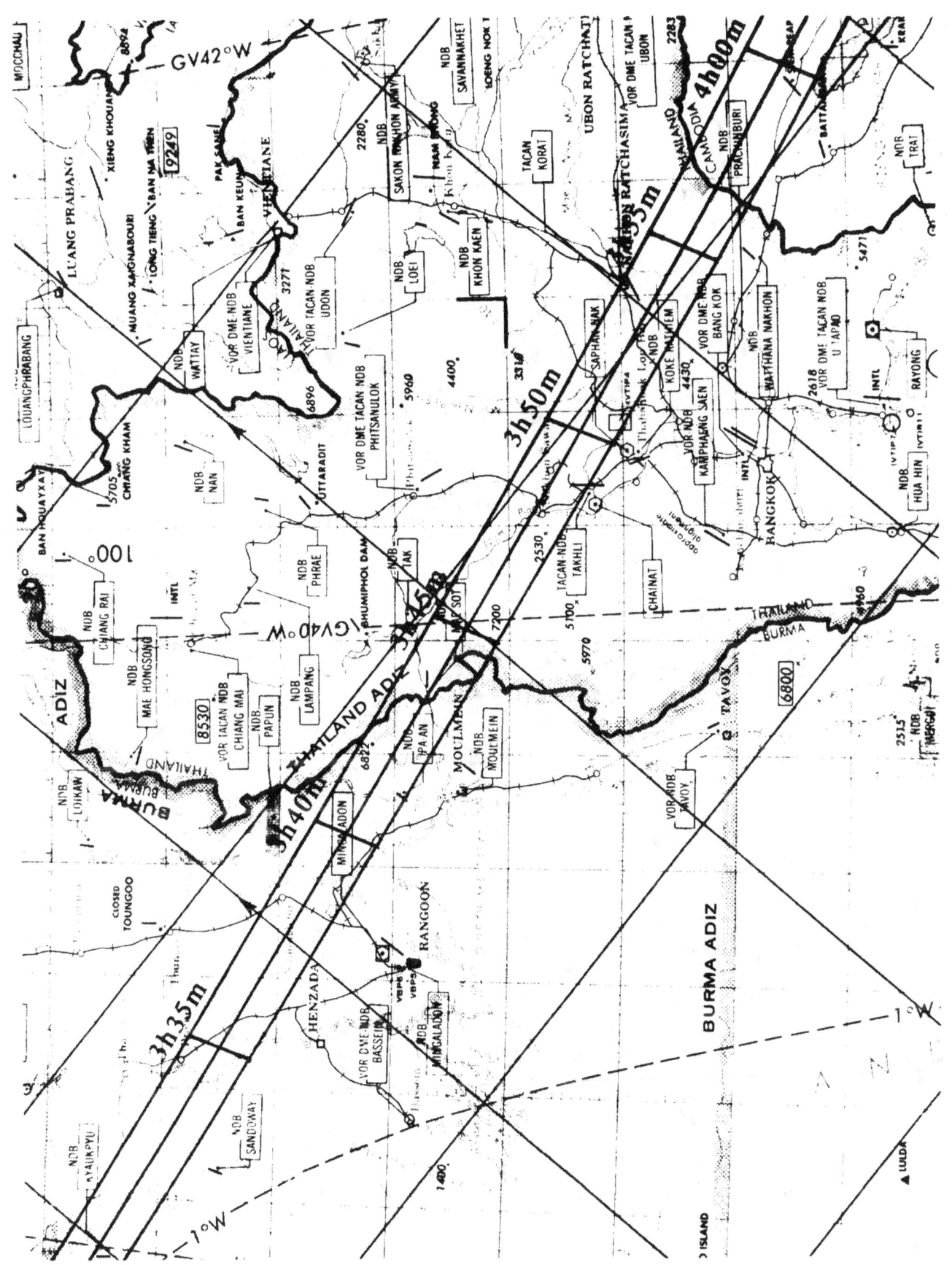

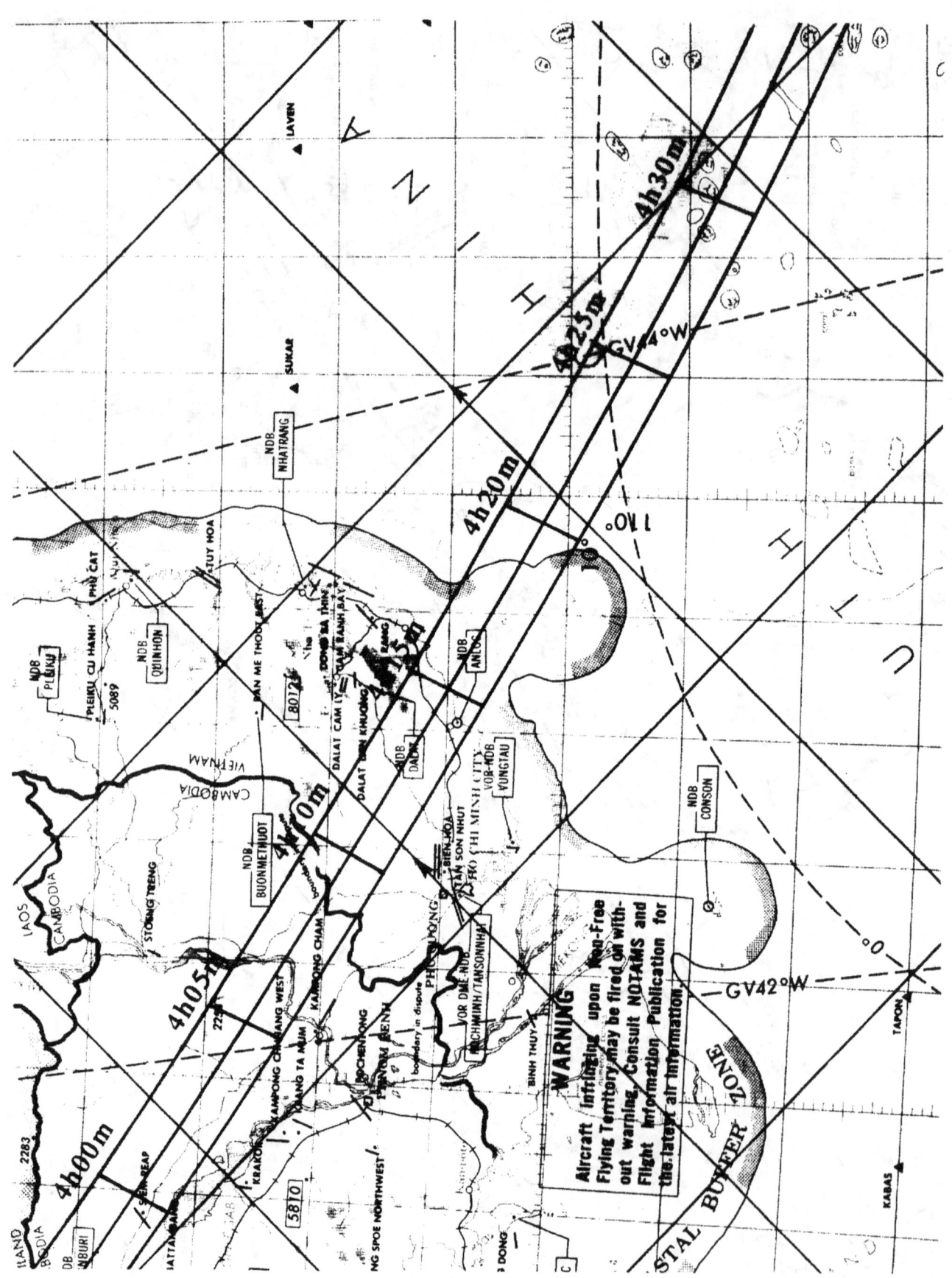

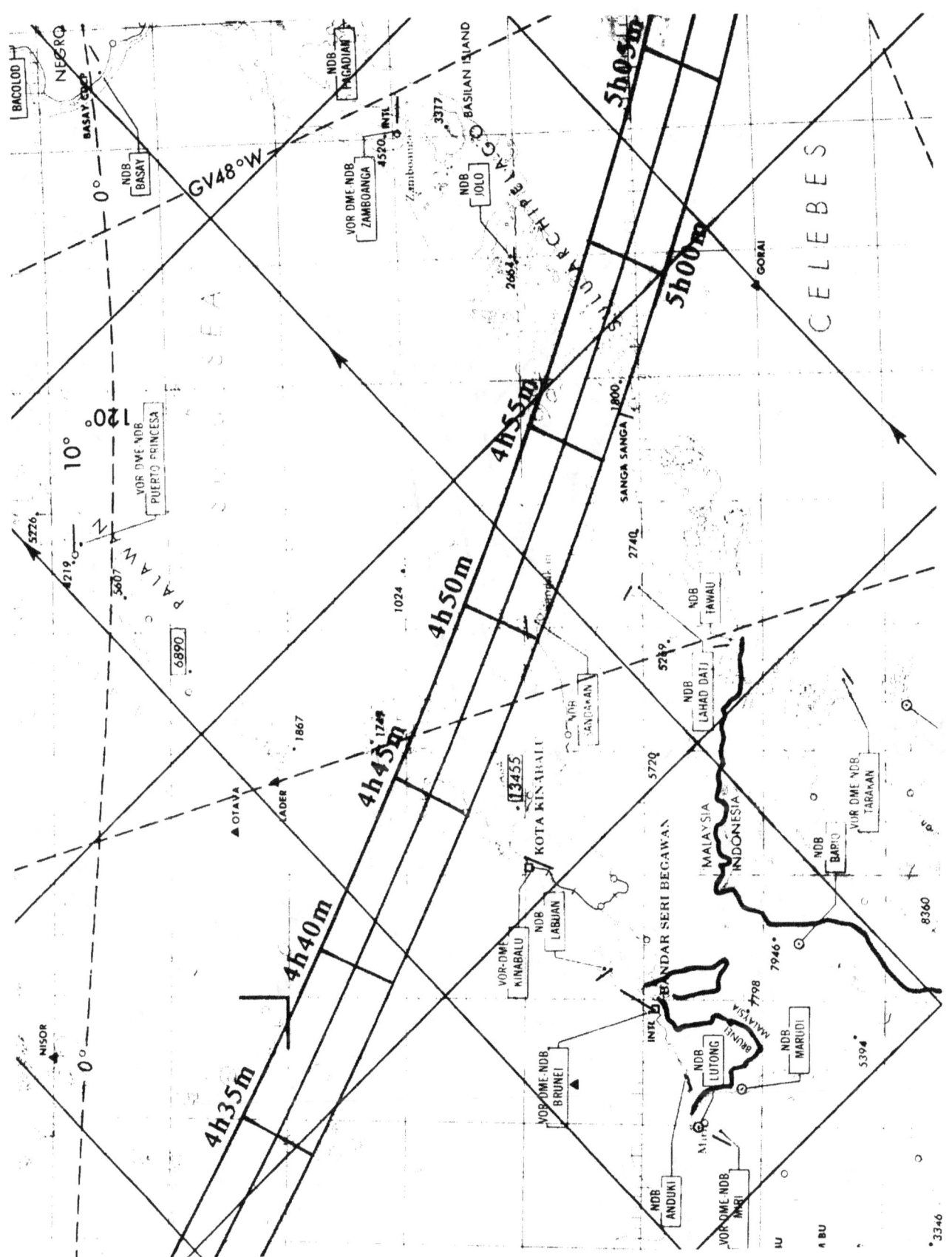

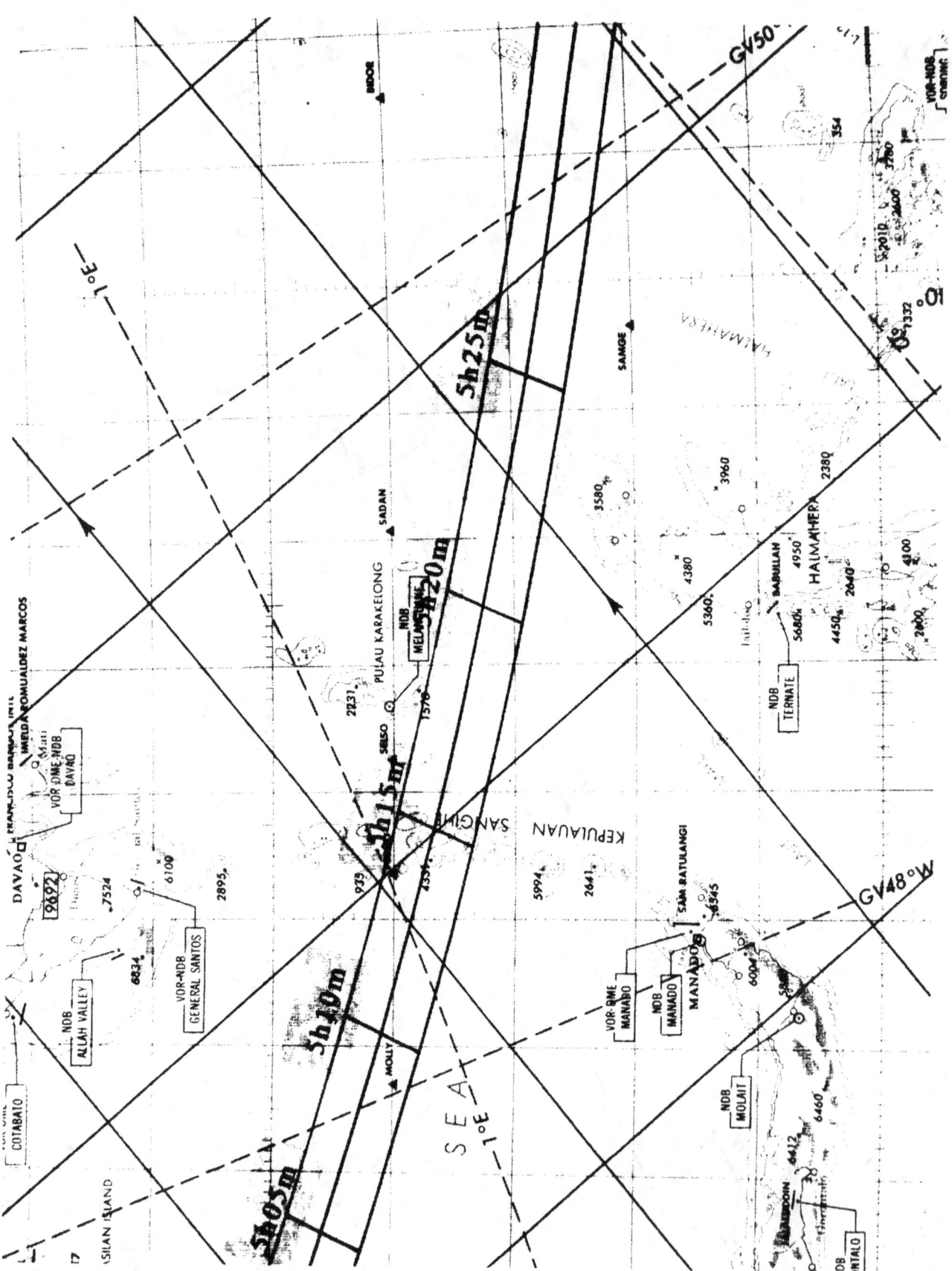

www.ingramcontent.com/pod-product-compliance
Lightning Source LLC
Chambersburg PA
CBHW081735170526
45167CB00009B/3833